GROUNDWATER CONTAMINATION FROM STORMWATER INFILTRATION

Robert Pitt

with contributions by

Shirley Clark, Keith Parmer, and Richard Field

Ann Arbor Press, Inc.
Chelsea, Michigan

Library of Congress Cataloging-in-Publication Data

Catalog record is available from the Library of Congress.

International Standard Book Number 1-57504-015-8

ANN ARBOR PRESS, INC.
121 South Main Street, Chelsea, Michigan 48118

Printed in the United States of America

1 2 3 4 5 6 7 8 9 0

ACKNOWLEDGMENTS

This book was prepared with the assistance of many individuals. Its creation would not have been possible without the diligent efforts of Shirley Clark, a Ph.D. student in the Department of Civil and Environmental Engineering at the University of Alabama who directed the literature search and organized much of the text. Keith Parmer, Research Assistant Professor in the Department of Civil and Environmental Engineering at the University of Alabama at Birmingham, provided valuable assistance in environmental chemistry. Many thanks are also due to Lyn Lewis, who typed the tables for the text.

Bob Kort, of the USDA, Winooski, Vermont, was very helpful in identifying many pieces of relevant literature. A number of University of Alabama at Birmingham Civil and Environmental Engineering students also volunteered their time to help with various aspects of this research. Shirita Scott and Camille Hubbard helped enter information into the bibliographic information system. Thanks are also extended to Patty Barron and Brian Robertson, who helped with many aspects in the preparation of this book.

The research to produce this book was supported by the U.S. EPA. Richard Field, Chief of the Storm and Combined Sewer Pollution Control Program, was the Project Officer and provided much valued direction. Michael Brown and Bill Vilkelis of his staff also provided important project assistance. Helpful comments from the report reviewers are also gratefully acknowledged.

ABOUT THE AUTHOR

 Bob Pitt received his Ph.D. in Civil and Environmental Engineering at the University of Wisconsin, his M.S.C.E. in Environmental Engineering and Hydraulic Engineering at San Jose State University, and his B.S. in Engineering Science from Humboldt State University. He is currently an Associate Professor in the Department of Civil and Environmental Engineering, with a secondary appointment in the Department of Environmental Health Sciences, at the University of Alabama at Birmingham. Prior to his teaching career, he was associated with a state regulatory agency and engineering consulting firms. He is a member of the American Society of Civil Engineers, the Water Environment Federation, the North American Lake Management Society, the Society for Environmental Toxicology and Chemistry, and the American Water Resources Association. He is a manuscript reviewer for numerous professional journals and reviews grant proposals for several state and national agencies. He has received several teaching awards and an award for distinguished achievement while at UAB. He has also received the top Alabama environmental volunteer award. He has published more than 40 journal articles, book chapters, and major research reports, and has made more than 50 national and international conference presentations in his career. He is a registered engineer in the State of Wisconsin and is a Diplomate of the American Academy of Environmental Engineers.

He currently teaches water supply and drainage design classes, hydrology and hydraulic engineering classes, experimental design and applied statistics classes, and stormwater management classes at UAB. He has also taught engineering professional development courses in stormwater management throughout North America, in Europe, and in Asia. His current research interests involve the development and testing of stormwater control practices at critical source areas, determining the level of control needed based on receiving water condition and exposures, developing more cost-effective monitoring strategies, and basin-wide stormwater management planning.

ABSTRACT

Many past research efforts have identified several organic and heavy metal toxicants in stormwater and surface receiving waters. Several projects have also documented the extensive nature of receiving water impacts from stormwater discharges to creeks and streams, especially biological impacts. Infiltration of stormwater has received much attention as a way to reduce these stormwater discharges to surface waters and to help restore the hydrologic balance in urban areas. The purpose of this book is to review the groundwater contamination literature as it relates to stormwater and wastewater contaminants known to exist in stormwater. This information is used to supplement the few studies that have examined groundwater contamination problems associated with stormwater infiltration. Potential problem pollutants are identified, based on their mobility, their abundance in stormwater, and their treatability before discharge. This information is used with published research results to identify the likely source areas for these potential problem pollutants. Recommendations are also made for stormwater infiltration guidelines in different areas and monitoring that should be conducted to evaluate a specific stormwater for groundwater contamination potential.

Contents

TABLES

Chapter 1

SUMMARY AND CONCLUSIONS

Prior to urbanization, groundwater recharge resulted from infiltration of precipitation through pervious surfaces, including grasslands and woods. This infiltrating water was relatively uncontaminated. With urbanization, the permeable soil surface area through which recharge by infiltration could occur was reduced. This resulted in much less groundwater recharge and greatly increased surface runoff. In addition, the waters available for recharge generally carried increased quantities of pollutants. With urbanization, new sources of groundwater recharge also occurred, including recharge from domestic septic tanks, percolation basins, and industrial waste injection wells, and from agricultural and residential irrigation. This book addresses potential groundwater problems associated with stormwater toxicants, and describes how conventional stormwater control practices can reduce these problems. This chapter is a summary of the main findings, conclusions, and recommendations contained in the main body of the book, where more detailed and referenced discussions are contained.

Chapter 2 presents a summary of the characteristics of urban runoff, especially from different source areas as monitored in an earlier phase of this EPA-funded research. This information is needed to identify critical sources of contaminants that may adversely affect ground and surface receiving waters. These sources can either be controlled, or otherwise diverted from the receiving waters. Chapter 3 is a literature review of potential groundwater impacts associated with pollutants that may be found in stormwater. This information is based on literature describing groundwater problem case studies associated with several classes of stormwater pollutants from different types of source waters, including: stormwater, sanitary wastewater, agricultural operations, and some industrial operations. Appendix A is an annotated bibliography of this literature, including abstracts of the references. The information in Chapters 2 and 3 can be used to identify which stormwater pollutants may adversely affect groundwater, and where they may originate. Chapter 4 summarizes the available treatment options for stormwater before discharge. This information can be used to suggest control methods that should be used at critical source areas, or at an outfall, before groundwater discharge. If pretreatment before infiltration is not practicable for a critical area, then other control options may need to be considered, especially pollution prevention or diversion of the runoff away from areas prone to groundwater infiltration.

This book presents information collected as part of a multi-year research project sponsored by the U.S. EPA. The first year of this research project included conducting the literature review and preparation of this book and the design and construction of a treatment device suitable for use at a class of critical source areas, specifically automobile service areas (gas stations, vehicle repair shops, public works maintenance yards, bus barns, etc.), and the installation of several inlet retrofitted devices to control runoff in residential and commercial areas. As part of this design effort, many samples were collected from a residential area, a school bus maintenance area, and two public works yards in Stafford, New Jersey. These samples were analyzed for a wide range of conventional and toxic pollutants that could potentially affect groundwater quality (if the water was infiltrated) and the design of the inlet devices and special treatment device. This information was used in the design of the critical area treatment device (the Multi-Chambered Treatment Tank, or MCTT) and will be included in an EPA research report.

The second project phase includes monitoring three inlet treatment devices and the MCTT stormwater treatment device over an extended period of time for many toxicants. Controlled tests for different filtering media are also being conducted during this project phase. These devices can all play a role in the pretreatment of stormwater before discharge to groundwaters. These research activities will result in design manuals for these devices for many U.S. locations. The third project phase is expected to include groundwater monitoring near an existing stormwater infiltrating trench gallery that has been operating in Stafford, New Jersey, for several years.

CHARACTERISTICS OF URBAN RUNOFF

Urban runoff is comprised of many different flow phases. These may include dry-weather base flows, stormwater runoff, combined sewer overflows (CSOs) and snowmelt. The relative magnitudes of these discharges vary considerably, based on a number of factors. Season (such as cold versus warm weather, or dry versus wet weather) and land use have been identified as important factors affecting base flow and stormwater runoff quality.

Land development increases stormwater runoff volumes and pollutant concentrations. Impervious surfaces, such as rooftops, driveways and roads, reduce infiltration of rainfall and runoff into the ground and degrade runoff quality. The most important hydraulic factors affecting urban runoff volume (and therefore the amount of water available for groundwater infiltration) are the quantity of rain and the extent of impervious surfaces directly connected to a stream or drainage system. Directly connected impervious areas include paved streets, driveways, and parking areas draining to curb and gutter drainage systems, and roofs draining directly to a storm or combined sewer pipe.

BOD$_5$ and nutrient concentrations in stormwater are lower than in raw sanitary wastewater; they are closer in quality to typically treated sanitary wastewaters. However, urban stormwater has relatively high concentrations of bacteria, along with high concentrations of many metallic and some organic toxicants.

Flow Phases

Possibly 25% of all separate stormwater outfalls have water flowing in them during dry weather, and as much as 10% are grossly contaminated with raw sewage, industrial wastewaters, etc. If stormwater is infiltrated before it enters the drainage system (such as by using French drains, infiltration trenches, grass swales, porous pavements or percolation ponds in upland areas) then the effects of inappropriate contaminated discharges into the drainage system on groundwater will be substantially reduced, compared to outfall infiltration practices. If outfall waters are to be infiltrated in larger regional facilities, then the effects of contaminated dry weather flows will have to be considered.

Pathogenic Microorganisms

Most of the effort in describing bacteria associated with urban runoff has involved fecal coliform analyses, mainly because of its historical use in water quality standards. However, many researchers have concluded that the fecal coliform test cannot be relied on to accurately assess the pathogenicity of recreational waters receiving urban runoff from uncontaminated storm sewers. Pathogenic bacteria have routinely been found in urban runoff at many locations.

Historically, fecal coliform standards of less than 200 organisms/100 mL have been recommended because the frequency of *Salmonella* detection has been found to increase sharply at fecal coliform concentrations greater than this value in waters receiving sanitary sewage discharges. The occurrence of salmonella biotypes in urban runoff is generally low, with their reported densities ranging from less than one to a high of ten organisms/100 mL. However, numerous urban runoff studies have not detected any *Salmonella*. In addition, *Salmonella* observations in urban runoff have not correlated well with fecal coliform observations. *Salmonella* is not usually considered a significant hazard in urban runoff because of the relatively large required infective dose and the low concentrations found in urban runoff.

The evidence of low densities and required high infective doses for *Salmonella* cannot minimize the health hazard of other pathogens that have been found in urban runoff (such as *P. aeruginosa*, *Shigella*, or enteroviruses) that do not require ingestion, or only require very low infective doses. *Shigella* species causing bacillary dysentery are one of the primary human enteric disease-producing bacteria present in water. *Pseudomonas* is reported to be the most abundant pathogenic bacteria organism in urban runoff and streams. Several thousand *Pseudomonas aeruginosa* organisms per 100 mL have been commonly found in many urban runoff samples. Relatively small populations of *P. aeruginosa* may be capable of causing water contact health problems ("swimmer's ear," and skin infections) and it is resistant to antibiotics. Pathogenic *E. coli* can also be commonly found in urban runoff.

Viruses may also be important pathogens in urban runoff. Very small amounts of a virus are capable of producing infections or diseases, especially when compared to the large numbers of bacterial organisms required for infection. Viruses are usually detected, but at low levels, in urban receiving waters and stormwater.

Toxicants

Nationwide testing has not indicated any significant regional differences in the toxicants detected, or in their concentrations. However, land use (especially residential versus industrial areas) has been found to be a significant factor in toxicant concentrations and yields. Concentrations of many urban runoff toxicants have exceeded the EPA water quality criteria for human health protection by large amounts.

Pesticides (α-BHC, γ-BHC, chlordane, and α-Endosulfan) are mostly found in dry-weather flows from residential areas, while heavy metals (As, Cd, Cr, Cu, Pb, and Zn) and other toxic materials (pentachlorophenol, bis (2-ethylhexyl) phthalate, and the PAHs: chrysene, fluoranthene, phenanthrene, and pyrene) are more prevalent in stormwater from industrial areas, although they are also commonly found in runoff from residential and commercial areas. Many of the heavy metals found in industrial area urban runoff are found at high concentrations during both dry weather and wet weather conditions.

Sources of Pollutants

High bacteria populations have been found in sidewalk, road, and some bare ground sheetflow samples (collected from locations where dogs would most likely be "walked"). Tables 1 and 2 summarize toxicant concentrations and likely sources or locations having some of the highest concentrations found during an earlier phase of this EPA funded research. The detection frequencies for the heavy metals are all close to 100% for all source areas, while the detection frequencies for the organics shown ranged from about 10 to 25%. Vehicle service areas had the greatest abundance of observed organics, with landscaped areas having many of the observed organics.

CONSTITUENTS OF CONCERN

Nutrients

Nitrates are one of the most frequently encountered contaminants in groundwater. Groundwater contamination of phosphorus has not been as widespread, or as severe, as that for nitrogen compounds.

Whenever nitrogen-containing compounds come into contact with soil, a potential for nitrate leaching into groundwater exists, especially in rapid-infiltration wastewater basins, stormwater infiltration devices, and agricultural areas. Nitrate has leached from fertilizers and affected groundwaters under various turf grasses in urban areas, including golf courses, parks, and home lawns. Significant leaching of nitrates occurs during the cool, wet seasons. Cool temperatures reduce denitrification and ammonia volatilization,

Table 1. Heavy Metal Source Area Observations

Toxicant	Highest Median Conc. (μg/L)	Source Area	Highest Conc. (μg/L)	Source Area
cadmium	8	vehicle service area runoff	220	street runoff
chromium	100	landscaped area runoff	510	roof runoff
copper	160	urban receiving water	1250	street runoff
lead	75	CSO	330	storage area runoff
nickel	40	parking area runoff	130	landscaped area runoff
zinc	100	roof runoff	1580	roof runoff

Table 2. Toxic Organic Source Area Observations

Toxicant	Maximum (μg/L)	Detection Frequency (%)	Significant Sources
benzo (a) anthracene	60	12	gasoline, wood preservative
benzo (b) fluoranthene	226	17	gasoline, motor oils
benzo (k) fluoranthene	221	17	gasoline, bitumen, oils
benzo (a) pyrene	300	17	asphalt, gasoline, oils
fluoranthene	128	23	oils, gasoline, wood preservative
naphthalene	296	13	coal tar, gasoline, insecticides
phenanthrene	69	10	oils, gasoline, coal tar
pyrene	102	19	oils, gasoline, bitumen, coal tar, wood preservative
chlordane	2.2	13	insecticide
butyl benzyl phthalate	128	12	plasticizer
bis (2-chloroethyl) ether	204	14	fumigant, solvents, insecticides, paints, lacquers, varnishes
bis (2-chloroisopropyl) ether	217	14	pesticides
1,3-dichlorobenzene	120	23	pesticides

and limit microbial nitrogen immobilization and plant uptake. The use of slow-release fertilizers is recommended in areas having potential groundwater nitrate problems. The slow-release fertilizers include urea formaldehyde (UF), methylene urea, isobutylidene diurea (IBDU), and sulfur-coated urea.

Residual nitrate concentrations are highly variable in soil due to differences in soil texture, mineralization, rainfall and irrigation patterns, organic matter content, crop yield, nitrogen fertilizer/sludge rate, denitrification, and soil compaction. Nitrate is highly soluble (>1 kg/L) and will stay in solution in the percolation water, after leaving the root zone, until it reaches the groundwater.

Pesticides

Urban pesticide contamination of groundwater can result from municipal and home-owner use of pesticides for pest control and their subsequent collection in stormwater runoff. Pesticides that have been found in urban groundwaters include: 2,4-D, 2,4,5-T, atrazine, chlordane, diazinon, ethion, malathion, methyl trithion, silvex, and simazine. Heavy repetitive use of mobile pesticides on irrigated and sandy soils likely contaminates groundwater. Fungicides and nematocides must be mobile in order to reach the target pest, and hence they generally have the highest contamination potential. Pesticide leaching depends on patterns of use, soil texture, total organic carbon content of the soil, pesticide persistence, and depth to the water table.

The greatest pesticide mobility occurs in areas with coarse-grained or sandy soils without a hardpan layer, having low clay and organic matter content and high permeability. Structural voids, which are generally found in the surface layer of finer-textured soils rich in clay, can transmit pesticides rapidly when the voids are filled with water and the adsorbing surfaces of the soil matrix are bypassed. In general, pesticides with low water solubilities, high octanol-water partitioning coefficients, and high carbon partitioning coefficients are less mobile. The slower-moving pesticides have been recommended in areas of groundwater contamination concern. These include the fungicides iprodione and triadimefon, the insecticides isofenphos and chlorpyrifos and the herbicide glyphosate. The most mobile pesticides include: 2,4-D, acenaphthylene, alachlor, atrazine, cyanazine, dacthal, diazinon, dicamba, malathion, and metolachlor.

Pesticides decompose in soil and water, but the total decomposition time can range from days to years. Literature half-lives for pesticides generally apply to surface soils and do not account for the reduced microbial activity found deep in the vadose zone. Pesticides with a thirty-day half-life can show considerable leaching. An order-of-magnitude difference in half-life results in a five- to tenfold difference in percolation loss. Organophosphate pesticides are less persistent than organochlorine pesticides, but they also are not strongly adsorbed by the sediment and are likely to leach into the vadose zone and the groundwater.

Other Organics

The most commonly occurring organic compounds that have been found in urban groundwaters include phthalate esters (especially bis (2-ethylhexyl) phthalate) and phenolic compounds. Other organics more rarely found, possibly due to losses during sample collection, have included the volatiles: benzene, chloroform, methylene chloride, trichloroethylene, tetrachloroethylene, toluene, and xylene. PAHs (especially benzo (a) anthracene, chrysene, anthracene and benzo (b) fluoroanthenene) have also been found in groundwaters near industrial sites.

Groundwater contamination from organics, like that from other pollutants, occurs more readily in areas with sandy soils and where the water table is near the land surface. Removal of organics from the soil and recharge water can occur by one of three methods: volatilization, sorption, and degradation. Volatilization can significantly reduce the concentrations of the most volatile compounds in groundwater, but the rate of gas transfer from the soil to the air is usually limited by the presence of soil water. Hydrophobic sorption onto soil organic matter limits the mobility of less soluble base/neutral and acid-extractable compounds through organic soils and the vadose zone. Sorption is not always a permanent removal mechanism, however. Organic resolubilization can occur during wet periods following dry periods. Many organics can be at least partially degraded by microorganisms, but others cannot. Temperature, pH, moisture content, ion exchange capacity of soil, and air availability may limit the microbial degradation potential for even the most degradable organic.

Pathogenic Microorganisms

Viruses have been detected in groundwater where stormwater recharge basins were located short distances above the aquifer. Enteric viruses are more resistant to environmental factors than enteric bacteria, and they exhibit longer survival times in natural waters. They can occur in potable and marine waters in the absence of fecal coliforms. Enteroviruses are also more resistant to commonly used disinfectants than are indicator bacteria, and can occur in groundwater in the absence of indicator bacteria.

The factors that affect the survival of enteric bacteria and viruses in the soil include pH, antagonism from soil microflora, moisture content, temperature, sunlight, and organic matter. The two most important attributes of viruses that permit their long-term survival in the environment are their structure and very small size. These characteristics permit virus occlusion and protection within colloid-size particles. Viral adsorption is promoted by increasing cation concentration, decreasing pH, and decreasing soluble organics. Since the movement of viruses through soil to groundwater occurs in the liquid phase and involves water movement and associated suspended virus particles, the distribution of viruses between the adsorbed and liquid phases determines the viral mass available for movement. Once the virus reaches the groundwater, it can travel laterally through the aquifer until it is either adsorbed or inactivated.

The major bacterial removal mechanisms in soil are straining at the soil surface and at intergrain contacts, sedimentation, sorption by soil particles, and inactivation. Because they are larger than viruses, most bacteria are therefore retained near the soil surface due to this straining effect. In general, enteric bacteria survive in soil from two to three months, although survival times of up to five years have been documented.

Heavy Metals and Other Inorganic Compounds

The heavy metals and other inorganic compounds in stormwater that are of most environmental concern, from a groundwater pollution standpoint, are aluminum, arsenic, cadmium, chromium, copper, iron, lead, mercury, nickel, and zinc. However, the majority of these compounds, with the consistent exception of zinc, are mostly found associated with the particulate solids in stormwaters and are thus relatively easily removed through sedimentation practices. Filterable forms of the metals may also be removed by sediment adsorption or organically complexed with other particulates.

In general, studies of recharge basins receiving large metal loads found that most of the heavy metals are removed either in the basin sediment or in the vadose zone. Dissolved metal ions are removed from stormwater during infiltration mostly by adsorption onto the near-surface particles in the vadose zone, while the particulate metals are filtered out at the soil surface. Studies at recharge basins found that lead, zinc, cadmium, and copper accumulated at the soil surface with little downward movement over many years. However, nickel, chromium, and zinc concentrations have exceeded regulatory limits in the soils below a recharge area at a commercial site. Elevated groundwater heavy metal concentrations of aluminum, cadmium, copper, chromium, lead, and zinc have been found below stormwater infiltration devices where the groundwater pH has been acidic. Allowing percolation ponds to go dry between storms can be counterproductive to the removal of lead from the water during recharge. Apparently, the adsorption bonds between the sediment and the metals can be weakened during the drying period.

Similarities in water quality between runoff water and groundwater have shown that there is significant downward movement of copper and iron in sandy and loamy soils. However, arsenic, nickel, and lead did not significantly move downward through the soil to the groundwater. The exception to this was some downward movement of lead with the percolation water in sandy soils beneath stormwater recharge basins. Zinc, which is more soluble than iron, has been found in higher concentrations in groundwater than iron. The order of attenuation in the vadose zone from infiltrating stormwater is: zinc (most mobile) > lead > cadmium > manganese > copper > iron > chromium > nickel > aluminum (least mobile).

Salts

The use of salt for winter traffic safety is a common practice in many northern areas, and the sodium and chloride, which are collected in the snowmelt, travel down through the vadose zone to the groundwater with little attenuation. Soil is not very effective at removing salts. Salts that are still in the percolation water after it travels through the vadose zone will contaminate the groundwater. Infiltration of stormwater has led to increases in sodium and chloride concentrations above background concentrations. Fertilizer and pesticide salts also accumulate in urban areas and can leach through the soil to the groundwater.

Studies of depth of pollutant penetration in soil have shown that sulfate and potassium concentrations decrease with depth, while sodium, calcium, bicarbonate, and chloride concentrations increase with depth. Once contamination with salts begins, the movement of salts into the groundwater can be rapid. The salt concentration may not decrease until the source of the salts is removed.

TREATMENT OF STORMWATER

Table 3 summarizes the filterable fraction of toxicants found in stormwater runoff sheet flows from many urban areas found during an earlier phase of this EPA-funded research. Pollutants that are mostly in filterable forms have a greater potential for affecting groundwater and are more difficult to control using conventional stormwater control practices that mostly rely on sedimentation and filtration principles. Luckily, most of the toxic organics and metals are associated with the nonfilterable (suspended solids) fraction of the stormwater. Likely exceptions include zinc, fluoranthene, pyrene,

Table 3. Reported Filterable Fractions of Stormwater Toxicants from Source Areas

Constituent	Filterable Fraction (%)
cadmium	20 to 50
chromium	<10
copper	<20
iron	small amount
lead	<20
nickel	small amount
zinc	>50
benzo (a) anthracene	none found in filtered fraction
fluoranthene	65
naphthalene	25
phenanthrene	none found in filtered fraction
pyrene	95
chlordane	none found in filtered fraction
butyl benzyl phthalate	irregular
bis (2-chloroethyl) ether	irregular
bis (2-chlorisopropyl) ether	none found in filtered fraction
1,3-dichlorobenzene	75

and 1,3-dichlorobenzene, which may be mostly found in the filtered sample portions. However, dry weather flows in storm drainage tend to have much more of the toxicants associated with filtered sample fractions.

Sedimentation is the most common fate and control mechanism for particulate related pollutants. This would be common for most stormwater pollutants, as noted above. Particulate removal can occur in many conventional stormwater control processes, including catchbasins, screens, drainage systems, and detention ponds. Sorption of pollutants onto solids and metal precipitation increase the sedimentation potential of these pollutants and also encourage more efficient bonding of the pollutants in soils, preventing their leaching to groundwaters. Detention ponds are probably the most common management practice for the control of stormwater runoff. If properly designed, constructed, and maintained, wet detention ponds can be very effective in controlling a wide range of pollutants. The monitored performance of wet detention ponds can be more than 90% removal of suspended solids, 70% for BOD_5 and COD, nutrient removals of about 60 to 70%, and heavy metal removals of about 60 to 95%. Catchbasins are very small sedimentation devices. Adequate cleaning can help to reduce the total solids and lead urban runoff yields by between 10 and 25%, and COD, total Kjeldahl nitrogen, total phosphorus, and zinc by between 5 and 10%. Other important fate mechanisms available in wet detention ponds, but which are probably not as important in small sump devices such as catchbasins, include volatilization and photolysis. Biodegradation, biotransformation, and bioaccumulation (into plants and animals) may also occur in ponds.

Upland infiltration devices (such as infiltration trenches, porous pavements, percolation ponds, and grass roadside drainage swales) are located at urban source areas. Infiltration (percolation) ponds are usually located at stormwater outfalls, or at large paved areas. These basins, along with perforated storm sewers, can infiltrate flows and pollutants from all upland sources combined. Infiltration devices can safely deliver large fractions of the surface flows to groundwater, if carefully designed and located. Local conditions that can make stormwater infiltration inappropriate include steep slopes, slowly percolating soils, shallow groundwater, and nearby groundwater uses. Grass filter strips may be quite effective in removing particulate pollutants from overland flows. The filtering effects of grasses, along with increased infiltration/recharge, reduce the particulate sediment load from urban landscaped areas. Grass swales are another type of infiltration device and can be used in place of concrete curb and gutters in most land uses, except possibly strip commercial and high-density residential areas. Grass swales allow the recharge of significant amounts of surface flows. Swales can also reduce the concentration of pollutants through filtration if the flows are shallow. Soluble and particulate heavy metal (copper, lead, zinc, and cadmium) concentrations can be reduced by at least 50%, and COD, nitrate nitrogen, and ammonia nitrogen concentrations can be reduced by about 25%, but only very small concentration reductions can be expected for organic nitrogen, phosphorus, and bacteria.

Sorption of pollutants to soils is probably the most significant fate mechanism of toxicants in biofiltration devices. Many of the devices also use sedimentation and filtration to remove the particulate forms of the pollutants from the water. Incorporation of the pollutants onto soil with subsequent biodegradation, with minimal resultant leaching to the groundwater, is desired. Volatilization, photolysis, biotransformation, and bioconcentration may also be significant in grass filter strips and grass swales. Underground French drains and porous pavements offer much less biological activity to reduce toxicants.

CONCLUSIONS

Table 4 is a summary of the pollutants found in stormwater that may cause groundwater contamination problems for various reasons. This table does not consider the risk associated with using groundwater contaminated with these pollutants. Causes of concern include high mobility (low sorption potential) in the vadose zone, high abundance (high concentrations and high detection frequencies) in stormwater, and high soluble fractions (small fraction associated with particulates that would have little removal potential using conventional stormwater sedimentation controls) in the stormwater. The contamination potential is the lowest rating of the influencing factors. As an example, if no pretreatment is to be used before percolation through surface soils, the mobility and abundance criteria are most important. If a compound is mobile, but is in low abundance (such as for VOCs), then the groundwater contamination potential will be low. However, if the compound is mobile and is also in high abundance (such as for sodium chloride, in certain conditions), then the groundwater contamination potential will be high. If sedimentation pretreatment is to be used before infiltration, then some of the pollutants will likely be removed before infiltration. In this case, all three influencing factors (mobility, abundance in stormwater, and soluble fraction) would be considered important. As an example, chlordane would have a low contamination potential with sedimentation pretreatment, while it would have a moderate contamination potential if no pretreatment were used. In addition, if subsurface infiltration/injection were used instead of surface percolation, the compounds would most likely be more mobile, making the abundance criteria the most important, with some regard given to the filterable fraction information for operational considerations.

Table 4. Groundwater Contamination Potential for Stormwater Pollutants

	Compounds	Mobility (Sandy/Low Organic Soils)	Abundance in Stormwater	Fraction Filterable	Contamination Potential for Surface Infiltration and No Pretreatment	Contamination Potential for Subsurface Infiltration with Sedimentation	Contamination Potential for Subsurface Injection with Minimal Pretreatment
Nutrients	nitrates	mobile	low/moderate	high	low/moderate	low/moderate	low/moderate
Pesticides	2,4-D	mobile	low	likely low	low	low	low
	γ-BHC (lindane)	intermediate	moderate	likely low	moderate	low	moderate
	malathion	mobile	low	likely low	low	low	low
	atrazine	mobile	low	likely low	low	low	low
	chlordane	intermediate	moderate	very low	moderate	low	moderate
	diazinon	mobile	low	likely low	low	low	low
Other organics	VOCs	mobile	low	very high	low	low	low
	1,3-dichloro-benzene	low	high	high	low	low	high
	anthracene	intermediate	low	moderate	low	low	low
	benzo (a) anthracene	intermediate	moderate	very low	moderate	low	moderate
	bis (2-ethylhexyl) phthalate	intermediate	moderate	likely low	moderate	low?	moderate
	butyl benzyl phthalate	low	low/moderate	moderate	low	low	low/moderate
	fluoranthene	intermediate	high	high	moderate	moderate	high
	fluorene	intermediate	low	likely low	low	low	low
	naphthalene	low/inter.	low	moderate	low	low	low
	pentachloro-phenol	intermediate	moderate	likely low	moderate	low?	moderate
	phenanthrene	intermediate	moderate	very low	moderate	low	moderate
	pyrene	intermediate	high	high	moderate	moderate	high

Table 4. Continued

	Compounds	Mobility (Sandy/Low Organic Soils)	Abundance in Stormwater	Fraction Filterable	Contamination Potential for Surface Infiltration and No Pretreatment	Contamination Potential for Subsurface Infiltration with Sedimentation	Contamination Potential for Subsurface Injection with Minimal Pretreatment
Pathogens	enteroviruses	mobile	likely present	high	high	high	high
	Shigella	low/inter.	likely present	moderate	low/moderate	low/moderate	high
	Pseudomonas aeruginosa	low/inter.	very high	moderate	low/moderate	low/moderate	high
	protozoa	low/inter.	likely present	moderate	low/moderate	low/moderate	high
Heavy metals	nickel	low	high	low	low	low	high
	cadmium	low	low	moderate	low	low	low
	chromium	inter./very low	moderate	very low	low/moderate	low	moderate
	lead	very low	moderate	very low	low	low	moderate
	zinc	low/very low	high	high	low	low	high
Salts	chloride	mobile	seasonally high	high	high	high	high

This table is only appropriate for initial estimates of contamination potential because of the simplifying assumptions made, such as the likely worst case mobility measures for sandy soils having low organic content. If the soil were clayey and had a high organic content, then most of the organic compounds would be less mobile than shown on this table. The abundance and filterable fraction information is generally applicable for warm weather stormwater runoff at residential and commercial area outfalls. The concentrations and detection frequencies would likely be greater for critical source areas (especially vehicle service areas) and critical land uses (especially manufacturing industrial areas).

The stormwater pollutants of most concern (those that may have the greatest adverse impacts on groundwaters) include:

- nutrients: nitrate has a low to moderate groundwater contamination potential for both surface percolation and subsurface infiltration/injection practices because of its relatively low concentrations found in most stormwaters. If the stormwater nitrate concentration were high, then the groundwater contamination potential would likely also be high.

- pesticides: lindane and chlordane have moderate groundwater contamination potentials for surface percolation practices (with no pretreatment) and for subsurface injection (with minimal pretreatment). The groundwater contamination potentials for both of these compounds would likely be substantially reduced with adequate sedimentation pretreatment.

- other organics: 1,3-dichlorobenzene may have a high groundwater contamination potential for subsurface infiltration/injection (with minimal pretreatment). However, it would likely have a lower groundwater contamination potential for most surface percolation practices because of its relatively strong sorption to vadose zone soils. Both pyrene and fluoranthene would also likely have high groundwater contamination potentials for subsurface infiltration/injection practices, but lower contamination potentials for surface percolation practices because of their more limited mobility through the unsaturated zone (vadose zone). Others (including benzo (a) anthracene, bis (2-ethylhexyl) phthalate, pentachlorophenol, and phenanthrene) may also have moderate groundwater contamination potentials, if surface percolation with no pretreatment or subsurface injection/infiltration is used. These compounds would have low groundwater contamination potentials if surface infiltration were used with sedimentation pretreatment. Volatile organic compounds (VOCs) may also have high groundwater contamination potentials if present in the stormwater (likely for some industrial and commercial facilities such as vehicle service establishments).

- pathogens: enteroviruses likely have a high groundwater contamination potential for all percolation practices and subsurface infiltration/injection practices, depending on their presence in stormwater (likely, especially if contaminated with sanitary sewage). Other pathogens, including *Shigella*, *Pseudomonas aeruginosa*, and various protozoa, will also have high groundwater contamination potentials if subsurface infiltration/injection practices are used without disinfection. If disinfection (especially by chlorine or ozone) is used, then disinfection byproducts (such as trihalomethanes or ozonated bromides) would have high groundwater contamination potentials.

- heavy metals: nickel and zinc would likely have high groundwater contamination potentials if subsurface infiltration/injection were used. Chromium and lead would have moderate groundwater contamination potentials for subsurface infiltration/

injection practices. All metals would likely have low groundwater contamination potentials if surface infiltration were used with sedimentation pretreatment.

- salts: chloride would likely have a high groundwater contamination potential in northern areas where road salts are used for traffic safety, irrespective of the pretreatment, infiltration, or percolation practice used.

Pesticides have been mostly found in urban runoff from residential areas, especially in dry-weather flows associated with landscaping irrigation runoff. The other organics, especially the volatiles, are mostly found in industrial areas. The phthalates are found in all areas. The PAHs are also found in runoff from all areas, but they are in higher concentrations and occur more frequently in industrial areas. Pathogens are most likely associated with sanitary sewage contamination of storm drainage systems, but several bacterial pathogens are also commonly found in surface runoff in residential areas. Zinc is mostly found in roof runoff and other areas where galvanized metal comes into contact with rainwater. Salts are at their greatest concentrations in snowmelt and early spring runoff in northern areas.

The control of these compounds will require a varied approach, including source area controls, end-of-pipe controls, and pollution prevention. All dry-weather flows should be diverted from infiltration devices because of their potentially high concentrations of soluble heavy metals, pesticides, and pathogens. Similarly, all runoff from manufacturing industrial areas should also be diverted from infiltration devices because of their relatively high concentrations of soluble toxicants. Combined sewer overflows should also be diverted because of sanitary sewage contamination. In areas of extensive snow and ice, winter snowmelt and early spring runoff should also be diverted from infiltration devices.

All other runoff should include pretreatment using sedimentation processes before infiltration, to both minimize groundwater contamination and to prolong the life of the infiltration device (if needed). This pretreatment can take the form of grass filters, sediment sumps, wet detention ponds, etc., depending on the runoff volume to be treated and other site specific factors. Pollution prevention can also play an important role in minimizing groundwater contamination problems, including minimizing the use of galvanized metals, pesticides, and fertilizers in critical areas. The use of specialized treatment devices can also play an important role in treating runoff from critical source areas before these more contaminated flows commingle with cleaner runoff from other areas. Sophisticated treatment schemes, especially the use of chemical processes or disinfection, may not be warranted, except in special cases, especially considering the potential of forming harmful treatment by-products (such as THMs and soluble aluminum).

The use of surface percolation devices (such as grass swales and percolation ponds), which have a substantial thickness of underlying soils above the groundwater, is preferable to using subsurface infiltration devices (such as dry wells, trenches or French drains, and especially injection wells), unless the runoff water is known to be relatively free of pollutants. Surface devices are able to take greater advantage of natural soil pollutant removal processes. However, unless all percolation devices are carefully designed and maintained, they may not function properly and may lead to premature hydraulic failure or contamination of the groundwater.

RECOMMENDATIONS

It has been suggested that, with a reasonable degree of site-specific design considerations to compensate for soil characteristics, infiltration can be very effective in controlling both urban runoff quality and quantity problems. This strategy encourages infiltration of urban runoff to replace the natural infiltration capacity lost through urbanization and to use the natural filtering and sorption capacity of soils to remove pollutants. However, potential groundwater contamination through infiltration of some types of urban runoff requires some restrictions. Infiltration of urban runoff having potentially high concentrations of pollutants that may pollute groundwater requires adequate pretreatment, or the diversion of these waters away from infiltration devices. The following general guidelines for the infiltration of stormwater and other storm drainage effluent are recommended in the absence of comprehensive site-specific evaluations:

- Runoff from residential areas (the largest component of urban runoff from most cities) is generally the least polluted urban runoff flow and should be considered for infiltration. Very little treatment of residential area stormwater runoff should be needed before infiltration, especially if surface infiltration is through the use of grass swales. If subsurface infiltration (French drains, infiltration trenches, dry wells, etc.) is used, then some pretreatment may be needed, such as the use of grass filter strips, or other surface filtration devices.
- Dry-weather storm drainage effluent should be diverted from infiltration devices because of the probable high concentrations of soluble heavy metals, pesticides, and pathogenic microorganisms.
- Combined sewage overflows should be diverted from infiltration devices because of their poor water quality, especially high pathogenic microorganism concentrations, and high clogging potential.
- Snowmelt runoff should also be diverted from infiltration devices because of its potential for having high concentrations of soluble salts.
- Runoff from manufacturing industrial areas should also be diverted from infiltration devices because of its potential for having high concentrations of soluble toxicants.
- Construction site runoff must be diverted from stormwater infiltration devices (especially subsurface devices) because of its high suspended solids concentrations, which would quickly clog infiltration devices.
- Runoff from other critical source areas, such as vehicle service facilities and large parking areas, should at least receive adequate pretreatment to eliminate their groundwater contamination potential before infiltration.

Recommended Stormwater Quality Monitoring to Evaluate Potential Groundwater Contamination

Most past stormwater quality monitoring has not been adequate to completely evaluate groundwater contamination potential. The following list shows the parameters that are recommended to be monitored if stormwater contamination potential needs to be considered, or infiltration devices are to be used. Other analyses are appropriate for additional monitoring objectives (such as evaluating surface water problems). In addition, all phases of urban runoff should be sampled, including stormwater runoff, dry-weather flows, and snowmelt.

- Contamination potential:
  — nutrients (especially nitrates)
  — salts (especially chloride)
  — VOCs (if expected in the runoff, such as from manufacturing industrial or vehicle service areas, could screen for VOCs with purgable organic carbon (POC) analyses)
  — pathogens (especially enteroviruses, if possible, along with other pathogens such as *Pseudomonas aeruginosa*, *Shigella*, and pathogenic protozoa)
  — bromide and total organic carbon (TOC) (to estimate disinfection by-product generation potential, if disinfection by either chlorination or ozone is being considered)
  — pesticides, in both filterable and total sample components (especially lindane and chlordane)
  — other organics, in both filterable and total sample components (especially 1,3-dichlorobenzene, pyrene, fluoranthene, benzo (a) anthracene, bis (2-ethyl-hexyl) phthalate, pentachlorophenol, and phenanthrene)
  — heavy metals, in both filterable and total sample components (especially chromium, lead, nickel, and zinc)

- Operational considerations:
  — sodium, calcium, and magnesium (in order to calculate the sodium adsorption ratio to predict clogging of clay soils)
  — suspended solids (to determine the need for sedimentation pretreatment to prevent clogging)

Chapter 2

CHARACTERISTICS OF URBAN RUNOFF

Unfortunately, some stormwaters from urban areas may be badly polluted. These waters may pose a potential threat to both surface and subsurface receiving waters. In order to protect these receiving water resources, treatment before discharge is likely needed. This chapter summarizes urban runoff quality. Many studies have investigated stormwater quality, with the EPA's Nationwide Urban Runoff Program (NURP) (EPA 1983) providing the largest and best known database. This water quality data will be compared to the information presented in Chapter 3 to identify which urban runoff waters need treatment to protect groundwaters.

Urban runoff is comprised of many different flow types. These include dry-weather base flows, stormwater runoff, combined sewer overflows (CSOs) and snowmelt. The relative magnitudes of these discharges vary considerably, based on a number of factors. Season (especially cold versus warm weather) and land use have been identified as important factors affecting base flow and stormwater runoff quality, respectively (Pitt and McLean 1986). This chapter briefly summarizes a number of different observations of runoff quality for these different phases and land uses, along with summaries of observations of source area flows contributing to these combined discharges. This information can be used to identify the best stormwater candidates for infiltration controls, and which ones to avoid.

STORMWATER CHARACTERISTICS

Land development increases stormwater pollutant concentrations and volumes. Impervious surfaces, such as rooftops, driveways and roads, reduce infiltration of rainfall and runoff into the ground and degrade runoff quality. Maintenance of landscaped areas further degrades runoff quality. The average runoff volume from developing subdivisions has been reported to be more than ten times greater than that of typical pre-development agricultural areas (Madison, et al. 1979).

Factors affecting runoff water volume (and therefore the amount of water available for groundwater infiltration) include rainfall quantity and intensity, slope, soil permeability, land cover, impervious area, and depression storage. Research during the Nationwide Urban Runoff Program (NURP) showed that the most important hydraulic factors affecting urban runoff volume were the quantity of rain and the extent of impervious surfaces directly connected to a stream or drainage system (EPA 1983). Directly connected impervious areas include paved streets, driveways, and parking areas draining to curb and gutter drainage systems, or roofs draining directly to a storm sewer.

Table 5 presents historical stormwater quality data (APWA 1969), while Table 6 is a summary of the Nationwide Urban Runoff Program stormwater data collected from about 1979 through 1982 (EPA 1983). BOD_5 and nutrient concentrations in stormwater are lower than for raw sanitary wastewater; they are closer in quality to typically treated sanitary wastewaters. However, urban stormwater has relatively high concentrations of bacteria, along with high concentrations of many metallic and some organic toxicants. As will be shown later, land use and source areas (parking areas, rooftops, streets, landscaped areas, etc.) all have important effects on stormwater runoff quality.

Urban Runoff Bacteria and Their Associated Public Health Significance

Most of the effort in describing bacteria characteristics of urban runoff has involved fecal coliform analyses, mainly because of its historical use in water quality standards. Fecal coliform bacteria observations have long been used as an indicator of sanitary sewage contamination and therefore has been used as an indicator of possible pathogenic microorganism contamination (Field and O'Shea 1992). Fecal streptococci analyses are also relatively common for urban runoff. Unfortunately, relatively few analyses of specific pathogenic microorganisms have been made for urban runoff.

The fecal coliform test is not specific for any one coliform type, or group of types, but instead has an excellent positive correlation for coliform bacteria derived from the intestinal tract of warm-blooded animals (Geldreich, et al. 1968). The fecal coliform test measures *Escherichia coli* as well as all other coliforms that can ferment lactose at 44.5°C and are found in warm-blooded fecal discharges. Geldreich (1976) found that the fecal coliform test represents over 96% of the coliforms derived from human feces and from 93 to 98% of those discharged in feces from other warm-blooded animals, including livestock, poultry, cats, dogs, and rodents. In many urban runoff studies, all of the fecal coliforms were *E. coli* (Qureshi and Dutka 1979). Field and O'Shea (1992) conclude that the fecal coliform test cannot be relied on to accurately assess the pathogenicity of recreational waters receiving urban runoff from uncontaminated storm sewers. The fecal streptococci test is sensitive to all of the intestinal *Streptococcus* bacteria from warm-blooded animal feces (Geldreich and Kenner 1969).

Pathogenic bacteria have been found in urban runoff at many locations and are probably from several different sources (Field, et al. 1976; Olivieri, et al. 1977; Qureshi and

Table 5. Characteristics of Stormwater Runoff

City	BOD₅ (mg/L)	Suspended Total Solids (mg/L)	Solids (mg/L)	Chlorides (mg/L)	COD (mg/L)
1. East Bay Sanitary District: Oakland, California					
Minimum	3		16	300	
Maximum	7,700	726	4,400	10,260	
Average	87	1,401	613	5,100	
2. Cincinnati, Ohio					
Maximum Seasonal Means	12				110
Average	17	260	227		111
3. Los Angeles County Average 1962–63	161	2,909		199	
4. Washington, D.C. Catch-basin samples during storm					
Minimum	6		26	11	
Maximum	625		36,250	160	
Average	126		2,100	42	
5. Seattle, Washington	10				
6. Oxney, England	100ᵃ	2,045			
7. Moscow, U.S.S.R.	186–285	1,000–3,500ᵃ			
8. Leningrad, U.S.S.R.	36	14,541			
9. Stockholm, Sweden	17–80	30–8,000			18–3,100
10. Pretoria, South Africa					
Residential	30				29
Business	34				28
11. Detroit, Michigan	96–234	310–914	102–213ᵇ		

ᵃMaximum; ᵇMean

Source: APWA 1969

Table 6. Median Stormwater Pollutant Concentrations for All Sites by Land Use (NURP)

Pollutant	Residential		Mixed Land Use		Commercial		Open/Nonurban	
	Median	COV[a]	Median	COV	Median	COV	Median	COV
BOD$_5$, mg/L	10.0	0.41	7.8	0.52	9.3	0.31	–	–
COD, mg/L	73	0.55	65	0.58	57	0.39	40	0.78
TSS, mg/L	101	0.96	67	1.14	69	0.85	70	2.92
Total Kjeldahl Nitrogen, µg/L	1900	0.73	1288	0.50	1179	0.43	965	1.00
NO$_2$ – N + NO$_3$ – N, µg/L	736	0.83	558	0.67	572	0.48	543	0.91
Total P, µg/L	383	0.69	263	0.75	201	0.67	121	1.66
Soluble P, µg/L	143	0.46	56	0.75	80	0.71	26	2.11
Total Lead, µg/L	144	0.75	114	1.35	104	0.68	30	1.52
Total Copper, µg/L	33	0.99	27	1.32	29	0.81	–	–
Total Zinc, µg/L	135	0.84	154	0.78	226	1.07	195	0.66

[a] COV: coefficient of variation = standard deviation/mean

Source: EPA 1983

Dutka 1979; Environment Canada 1980; Pitt 1983; Pitt and McLean 1986; and Field and O'Shea 1992). Table 7 summarizes the occurrence of various pathogenic bacteria types found in urban stormwaters at Burlington, Ontario; Milwaukee; and Cincinnati. The observed ranges of concentrations and percentage isolations of these biotypes vary significantly from site to site and at the same location for different times. However, it is seen that many of the potentially pathogenic bacteria biotypes can be present in urban stormwater runoff. The occurrence of *Salmonella* biotypes is generally low and their reported density is usually less than one organism/100 mL. *Pseudomonas aeruginosa* are frequently encountered at densities of hundreds to thousands of organisms/100 mL.

Some authors do not feel that urban runoff presents a significant health problem. Olivieri, et al. (1977) do not believe that urban runoff constitutes a major health problem because of the large numbers of viable bacteria cells that must be consumed to establish an infection for many of the pathogens found in urban runoff. For urban runoff, it may be impossible to consume enough bacteria cells to establish the infective dose. The importance of urban runoff in disease transmission in the Ottawa area was also questioned by Gore and Storrie/Proctor and Redfern (1981). They stated that little or no correlation was found between fecal coliform indicator bacteria and pathogenic bacteria in Ottawa stormwater runoff and local receiving waters. They further stated that the currently applied objectives in Ontario for fecal coliforms for body contact recreation are neither universal nor absolute standards relating to disease or infection. They concluded that these numeric objectives should be reviewed for their applicability to the local swimming beaches. However, Field and O'Shea (1992) pointed out that the evidence of low densities and required high infective doses for some pathogens cannot minimize the health hazard of other pathogens that have been found in urban runoff (such as *P. aeruginosa*, *Salmonella typhosa*, *Shigella*, or enteroviruses) that do not require ingestion, or only require very low infective doses.

Urban Runoff Salmonella *Observations*

Salmonella has been reported in some, but not all, urban stormwaters. Qureshi and Dutka (1979) frequently detected *Salmonella* in southern Ontario stormwaters. They did not find any predictable patterns of *Salmonella* isolations, as they were found throughout the various sampling periods. Olivieri, et al. (1977) frequently found *Salmonella* in Baltimore runoff, but at relatively low concentrations. Typical concentrations were from 5 to 300 *Salmonella* organisms per ten liters. The concentrations of *Salmonella* were about ten times higher in the Baltimore stormwater samples than in the urban stream receiving the runoff. They also did not find any marked seasonal variations in *Salmonella* concentrations. Almost all of the stormwater samples that had fecal coliform concentrations greater that 2000 organisms/100 mL had detectable *Salmonella* concentrations. However, about 27% of the samples having fecal coliform concentrations less than 200 organisms/100 mL had detectable *Salmonella*.

Quite a few urban runoff studies did not detect *Salmonella*. Schillinger and Stuart (1978) found that *Salmonella* isolations were not common in a Montana subdivision study and that the isolations did not correlate well with fecal coliform concentrations. Environment Canada (1980) stated that *Salmonella* were virtually absent from Ottawa storm drainage samples obtained in 1979. They concluded that *Salmonella* are seldom

Table 7. Pathogenic Microorganisms Found in Urban Stormwater (organisms/100 mL)

City, Province/State	Catchment/Land Use	Staphylococcus aureus	Pseudomonas aeruginosa	Salmonella	Streptococci	Reference
Burlington, Ontario	Aldershot Plaza		14–3,000	S. seftenberg & S. newport isolated		Qureshi and Dutka, 1979
Malvern Road			<1–740	100% negative		Qureshi and Dutka, 1979
Milwaukee, WI	highway runoff	all <1,000	all <1,000	45% positive		Gupta, et al, 1981
Cincinnati, OH	business district				79% positive[a]	Geldreich and Kenner, 1969
	residential area				80% positive[b]	Geldreich and Kenner, 1969
	rural area				87% positive[c]	Geldreich and Kenner, 1969

[a] *Strep.* bacteria types found: *S. bovis/S. equinus* (2%)
   Atypical *S. faecalis* (1%)
   S. faecalis liquifaciens (18%)
   S. thompson: 4,500/100 mL

[b] *Strep.* bacteria types found: *S. bovis/S. equinus* (0.5%)
   Atypical *S. faecalis* (1%)
   S. faecalis liquifaciens (18%)

[c] *Strep.* bacteria types found: *S. bovis/S. equinus* (0.5%)
   Atypical *S. faecalis* (0.2%)
   S. faecalis liquifaciens (12%)

present in significant numbers in Ottawa urban runoff. The types of *Salmonella* found in southern Ontario were *S. thompson* and *S. typhimurium* var. *copenhagen* (Qureshi and Dutka 1979).

Olivieri, et al. (1977) stated that the primary human enteric disease–producing *Salmonella* biotypes associated with the ingestion of water include *S. typhi* (typhoid fever), *S. paratyphi* (paratyphoid fever), and *Salmonella* species (salmonellosis). These biotypes are all rare, except for *Salmonella*. The dose of *Salmonella* required to produce an infection in a healthy adult is quite large (approximately 100,000 organisms). However, more sensitive individuals, such as children and the elderly, are much more susceptible to disease. The salmonellosis health hazard associated with urban streams is believed to be small because of this relatively large infective dose. If two liters of stormwater having typical *Salmonella* concentrations (10 organisms/10 L) are ingested, less than 0.001 of the required infective dose would be ingested. If a worst-case *Salmonella* stormwater concentration of 10,000 organisms/10 L occurred, the ingestion of 20 L of stormwater would be necessary for an infective dose. They stated that the low concentrations of *Salmonella,* coupled with the unlikely event of consuming enough stormwater, make the *Salmonella* health hazard associated with urban runoff small.

Geldreich (1965) recommended a fecal coliform standard of 200 organisms/100 mL because the frequency of *Salmonella* detection increased sharply at fecal coliform concentrations greater than this value. Setmire and Bradford (1980) stated that the National Academy of Sciences recommended a fecal coliform standard of 70/100 mL in waters with shellfish harvesting to restrict *Salmonella* concentrations in edible tissues. However, Field, et al. (1976) concluded that the use of indicator bacteria to predict *Salmonella* ingestion is less meaningful in stormwater runoff than in other waters.

Urban Runoff Shigella Observations

Olivieri, et al. (1977) stated that there is circumstantial evidence that *Shigella* is present in urban runoff and receiving waters and could present a significant health hazard. There have been problems in isolating and quantifying *Shigella* bacteria. *Shigella* species causing bacillary dysentery are one of the primary human enteric disease-producing bacteria agents present in water. The infective dose of *Shigella* necessary to cause dysentery is quite low (10 to 100 organisms). Because of this low required infective dose and the assumed presence of *Shigella* in urban waters, it may be a significant health hazard associated with urban runoff.

Urban Runoff Pseudomonas aeruginosa Observations

Pseudomonas is reported to be the most abundant pathogenic bacteria organism in urban runoff and streams (Olivieri, et al. 1977). Pitt and McLean (1986) found *P. aeruginosa* populations of several thousand organisms per 100 mL in many urban runoff samples. No information could be found concerning the problems associated with ingestion of *P. aeruginosa*–contaminated drinking waters. However, relatively small populations of *P. aeruginosa* may be capable of causing water contact health problems ("swimmer's ear," and skin infections) and it is resistant to antibiotics.

Other Urban Runoff Pathogen Observations

E. coli and *Vibrio cholerae* are disease-producing pathogens associated with the ingestion of water. The cholera pathogen is quite rare, but *E. coli* is more common in urban runoff. The required infective oral dose of both of these pathogens is about 10^8 organisms for healthy adults (Olivieri, et al. 1977).

Viruses may also be significant pathogens in urban runoff. Very small amounts of a virus are capable of producing infections or diseases, especially when compared to the large numbers of bacteria organisms required for infection (Berg 1965). Olivieri, et al. (1977) stated that viruses are usually detected at low levels in urban receiving waters and storm runoff.

Summary of Urban Runoff Pathogenic Microorganism Observations

Many potentially pathogenic bacteria biotypes may be present in urban runoff. Because of the low probability of direct ingestion of urban runoff, many of the potential human diseases associated with these biotypes are not likely to occur in normal receiving waters. The pathogenic organisms of most concern in urban runoff (and therefore that have received the most attention) are usually associated with skin infections and body contact in recreational waters. The most significant stormwater biotype causing skin infections would be *Pseudomonas aeruginosa.* This biotype has been detected frequently in most urban runoff studies in concentrations that may cause potential infections. *Shigella* may be present in urban runoff and receiving waters. This pathogen, when ingested in low numbers, can cause dysentery. A number of other pathogenic microorganisms have been periodically reported in urban runoff.

Significant Stormwater Toxicants

Stormwater research has quantified some inorganic and organic hazardous and toxic substances frequently found in urban runoff. The NURP data (Table 8), collected from mostly residential areas throughout the U.S., did not indicate any significant regional differences in the substances detected, or in their concentrations (EPA 1983). However, the residential and industrial data obtained by Pitt and McLean (1986) in Toronto found significant concentration and yield differences for these two distinct land uses and for dry-weather and wet-weather urban runoff flows.

The concentrations of many of these toxic pollutants exceeded the U.S. EPA water quality criteria for human health protection by large amounts. As an example, typical standards for PAHs in surface waters used as drinking water supplies are 2.8 ng/L (0.0028 μg/L) (EPA 1986). As shown in Table 8, urban runoff concentrations of chrysene (0.6 to 10 μg/L), fluoranthene (0.3 to 21 μg/L), phenanthrene (0.3 to 10 μg/L) and pyrene (0.3 to 16 μg/L) (four of the most common PAHs found in urban runoff) have been reported to be from 100 to as much as several thousand times greater than this criterion.

Table 9 lists the toxic and hazardous substances that have been found in more than 10% of the industrial and residential urban runoff samples analyzed (Galvin and Moore

Table 8. Summary of NURP Priority Pollutant Analyses[a] (Only those compounds found in greater than 10% of outfall samples are shown)

Substance	Frequency of Detection %	Range of Detected Concentrations (µg/L)
Pesticide		
α - BHC	20	0.0027 to 0.1
γ - BHC (lindane)	15	0.007 to 0.1
chlordane	17	0.01 to 10
α - endosulfan	19	0.008 to 0.2
Metals and Cyanide		
antimony	13	2.6 to 23
arsenic	52	1 to 51
beryllium	12	1 to 49
cadmium	48	0.1 to 14
chromium	58	1 to 190
copper	91	1 to 100
cyanides	23	2 to 300
lead	94	6 to 460
mercury	10	0.6 to 1.2
nickel	43	1 to 182
selenium	11	2 to 77
zinc	94	10 to 2400
PCBs and Related Compounds (none detected in greater than 1% of all samples)		
Halogenated Aliphatics		
methylene chloride	11	5 to 15
Ethers (none detected in any of the samples)		
Monocyclic Aromatics (none detected in greater than 6% of all samples)		
Phenols and Cresols		
phenol	14	1 to 13
pentachlorophenol	19	1 to 115
4-nitrophenol	10	1 to 37
Phthalate Esters		
bis (2-ethylhexyl) phthalate	22	4 to 62
Polycyclic Aromatic Hydrocarbons		
chrysene	10	0.6 to 10
fluoranthene	16	0.3 to 21
phenanthrene	12	0.3 to 10
pyrene	15	0.3 to 16

[a] Based on 121 samples from 17 cities

Source: EPA 1983.

Table 9. Hazardous and Toxic Substances Found in Urban Runoff[a]

	Residential Areas	Industrial Areas
Halogenated Aliphatics		
1,2,-dichlorethene		x
methylene chloride		x
tetrachloroethylene		x
Phthalate Esters		
bis (2-ethylene) phthalate	x	
butyl benzyl phthalate	x	x
diethyl phthalate		x
di-n-butyl phthalate	x	x
Polycyclic Aromatic Hydrocarbons		
phenanthrene		x
pyrene		x
Other Volatiles		
benzene	x	x
chloroform		x
ethylbenzene		x
n-nitro-sodimethylamine		x
toluene		x
Heavy Metals		
aluminum	x	x
chromium		x
copper	x	x
lead	x	x
zinc	x	x
Pesticides and Phenols		
BHC	x	
chlordane	x	
dieldrin	x	
endosulfan sulfate	x	
endrin	x	
isophorone	x	
methoxychlor	x	
PCB-1254		x
PCB-1260		x
pentachlorophenol	x	x
phenol	x	x

[a] Substances found in at least 10% of the stormwater samples analyzed.

Sources: Galvin and Moore (1982), EPA (1983), and Pitt and McLean (1986).

1982; EPA 1983; and Pitt and McLean 1986). As noted above, available NURP data do not reveal toxic urban runoff conditions significantly different for different parts of the U.S. (EPA 1983). The pesticides shown were mostly found in urban runoff from residential areas, while heavy metals and other hazardous materials were much more prevalent in industrial areas. Urban runoff dry-weather base flows may also be significant contributors of hazardous and toxic pollutants. Lindane (gamma-BHC) and dieldrin may be common in residential dry-weather storm sewer flows, while PCBs may be common in industrial dry-weather storm sewer flows. Many of the heavy metals found in industrial area urban runoff were found at high concentrations during both dry-weather and wet-weather conditions.

COMBINED SEWAGE CHARACTERISTICS

Combined sewage is made up of raw sanitary wastewater alone during dry-weather conditions, but also includes stormwater during wet-weather conditions. Because of the relatively slow sewage flow rates during dry weather, many combined sewage systems experience deposition of solids in the sewerage system. When stormwater enters the system during wet weather, this deposited material is flushed from the system. This "first flush" therefore typically has greater pollutant concentrations than either separate stormwater or separate sanitary sewage (Moffa 1989). Tables 10 through 14 summarize various characteristics of combined sewage. Table 12 compares bacteria densities of combined sewage with separate stormwater in Baltimore (Olivieri, et al. 1977). This table shows very little difference in the bacteria densities of these two sample types. Tables 13 and 14 show heavy metal and other toxicant concentrations in combined sewage. Pitt and Barron (1990) found all of the heavy metals investigated in New York City combined sewage samples, but only two of the base-neutral organic compounds were found in more than one of the 20 samples analyzed (di-n-butyl phthalate and bis (2-ethyl hexyl) phthalate). None of the base-neutrals were detected (at a detection limit of about 1 μg/L) in the filtered sample portions, but most of the metals were found in the filtered samples.

RELATIVE CONTRIBUTIONS OF URBAN RUNOFF FLOW PHASES

Tables 15 and 16 summarize Toronto residential/commercial and industrial urban runoff characteristics during both warm and cold weather (Pitt and McLean 1986). These tables show the relative importance of wet-weather and dry-weather flows coming from separate stormwater systems. If urban runoff is to be directed to an outfall infiltration device, then the dry-weather flows will also be present at the outfalls. Possibly 25% of all separate stormwater outfalls have water flowing in them during dry weather, and as many as 10% are grossly contaminated with raw sewage, industrial wastewaters, etc. (Pitt, et al. 1993). The EPA's Stormwater Permit program requires municipalities to conduct stormwater outfall surveys to identify, and then correct, inappropriate discharges into separate storm drainage. However, it can be expected that substantial outfall contamination will exist for many years. If stormwater is infiltrated before it enters the drainage system (e.g., by using French drains, infiltration trenches, grass swales, porous pavements, or percolation ponds in upland areas) then the effects of contamination problems in the drainage system on groundwater will be substantially reduced. If outfall waters are to be infiltrated in larger regional facilities, then these periods of dry-weather flows will have to be considered.

Similar problems occur in areas having substantial snowfalls. Table 16 summarizes Toronto snowmelt and cold weather baseflow characteristics (Pitt and McLean 1986).

Table 10. Pollutant Concentrations in Combined Sewer Overflows

					Average Pollutant Concentration (mg/L)					
	TSS	VSS	BOD$_5$	COD	Kjeldahl Nitrogen	Total Nitrogen	PO$_4$-P	OPO$_4$-P	Lead	Fecal Coliforms[a]
Des Moines, Iowa	413	117	64	—	—	4.3	1.86	1.31	—	—
Milwaukee, Wisconsin	321	109	59	264	4.9	6.3	1.23	0.86	—	—
New York City, New York										
Newtown Creek	306	182	222	481	—	—	—	—	—	—
Spring Creek	347	—	111	358	—	16.6	4.5[b]	—	0.60	—
Poissy, France[c]	751	387	279	1005	—	43	17[b]	—	—	—
Racine, Wisconsin	551	154	158	—	—	—	2.78	0.92	—	201
Rochester, New York	273	—	65	—	2.6	—	—	0.88	0.14	1140
Average (not weighted)	370	140	115	367	3.8	9.1	1.95	1.00	0.37	670
Range	273–551	109–182	59–222	264–481	2.6–4.9	4.3–16.6	1.23–2.78	0.86–1.31	0.14–0.60	201–1140
Syracuse, New York[d]	306	98.4	64.3	—	2.83	—	0.35[b]	—	—	1,407
Hartford, Connecticut[e]	727	—	33	319	6.20(NH$_3$)	—	—	—	—	2,600

[a] 1000 organisms/100 mL.

[b] Total P (not included in average).

[c] Not included in average because of high strength of municipal sewage when compared to the United States.

[d] O'Brien & Gere, 1979. "Combined Sewer Overflow Abatement Program," a report to Onondaga County, NY.

[e] Calocerinos & Spina, 1988. "Hartford Combined Sewer Overflow Facility Plan," a report to the Metropolitan District Commission, Hartford, CT.

Ref. EPA 600/8-77-014 *Urban Stormwater Management and Technology, Update and Users Guide.*

Source: Moffa 1989.

Table 11. Selected Combined Sewer Overflow Bacteria Data (organisms/100 mL)

City (reference)	Fecal Coliforms	Fecal Streptococci
Ottawa (Ontario Ministry of the Environment 1983)	5×10^5–9×10^6	—
Toronto (Ontario Ministry of the Environment 1982)	10^6	—
Detroit (Geldreich 1976)	10^6–10^7	10^5
Selected Nationwide Data (Field, et al. 1972)	2×10^4–2×10^7	2×10^4–2×10^6

The bacteria densities during cold weather are substantially less than during warm weather, but are still relatively high (EPA 1983). However, chloride concentrations and dissolved solids are much higher during cold weather. Early spring stormwater events also contain high dissolved solids concentrations (Bannerman, personal communication, WI DNR). Fortunately, upland infiltration devices do not work well during cold weather due to freezing soils, possibly minimizing groundwater contamination. However, outfall flows occur under ice into receiving waters (including detention ponds) and may enter regional infiltration devices if not specifically diverted.

POLLUTANT CONTRIBUTIONS FROM DIFFERENT URBAN SOURCE AREAS

Limited source area sheetflow quality data are available from several studies conducted in California, Washington, Nevada, Wisconsin, Illinois, Ontario, Colorado, New Hampshire, New York, and Alabama since 1979. A relatively large amount of parking and roof runoff quality data has been obtained from all of these locations, but only a few of these studies evaluated a broad range of source areas or land uses. This information can be used to identify which upland areas can be controlled by direct infiltration, which ones would require significant pretreatment before infiltration, and which ones should not be discharged to the groundwater because of potentially high concentrations of problem pollutants that may not be able to be adequately treated. The major urban source area categories that have been studied include:

- roofs
- paved parking areas
- paved storage areas
- unpaved parking and storage areas
- driveways
- streets
- landscaped areas
- undeveloped areas
- freeway paved lanes and shoulders
- vehicle service areas

Table 12. Geometric Mean Densities of Selected Pathogens and Indicator Microorganisms in Baltimore Stormwater and Combined Sewage

Sampling Station	Sample Type	Enterovirus PFU/10 L	Salmonella sp. MPN/10 L	P. aeruginosa MPN/10 L	Staph. aureus MPN/10 mL ($\times 10^4$)	TC MPN/100 ($\times 10^3$)	FC MPN/100 mL ($\times 10^4$)	FS #/100 mL ($\times 10^4$)	Enterococci #/100 mL
Stoney Run	combined	190	30	1300	12	4.8	19	4.1	1.4
Glen Avenue	combined	75	24	3300	14	24	81	66	21
Howard Park	combined	280	140	5200	36	120	450	24	5.9
Jones Falls	combined	30	25	6600	40	29	120	28	8.7
Bush Street	stormwater	6.9	30	2000	120	38	83	56	12
Northwood	stormwater	170	5.7	590	12	3.8	6.9	5	2.1

Source: Olivieri, et al. 1977.

Table 13. Concentrations of Heavy Metals in Combined Sewer Overflows

Land Use Type	Copper (μg/L)	Zinc (μg/L)	Lead (μg/L)
Medium-Density Residential[a]	77	191	93
High-Density Residential[b]	48	185	84
Residential/Commercial	100	255	135
Light Industrial	58	136	47
Heavy Industrial	98	447	223
Mean for All Land Uses	76	242	116

[a] Medium-Density Residential: 3 to 8 dwelling units per acre.

[b] High-Density Residential: 9 or more dwelling units per acre.

Source: Johnson 1990.

Tables 17 through 20 summarize much of the data available describing urban area runoff pollutants from these source areas for different land uses and seasons.

Lead and zinc concentrations are generally the highest in sheetflows from paved parking areas and streets, with some high zinc concentrations also found in roof drainage samples. High bacteria populations have been found in sidewalk, road, and some bare ground sheetflow samples (collected from locations where dogs would most likely be "walked").

Pentachlorophenol was detected (400 to 500 ng/L concentrations) in four of the five industrial source area samples analyzed for priority pollutants in Toronto (Pitt and McLean 1986). These samples were collected from an industrial subdrainage area and from a paved storage yard. Two of the five industrial source area priority pollutant sheetflow samples analyzed also had detectable PCBs (80 and 190 ng/L), α-BHC (8 and 10 ng/L), and γ-BHC (2 and 10 ng/L) concentrations.

Some of the sheetflow contributions observed at these locations were not sufficient to explain the concurrent concentrations observed in runoff at outfalls. The low chromium surface sheetflow concentrations and the high outfall concentrations at the Toronto industrial area, as an example, indicated a high likelihood of direct industrial wastewater connections to the storm drainage system. Chromium was rarely detected in any sheetflow samples, but was commonly found in potentially problem concentrations at the industrial outfall. Similarly, most of the fecal coliform populations observed in sheetflows were significantly lower than those observed at the outfall.

The following paragraphs briefly summarize likely sources of important pollutants in urban areas:

Cadmium was commonly detected by Pitt, et al. (1995) in almost all stormwater source area runoff samples. They found the highest median concentration (8 μg/L) in vehicle service area runoff, while a street runoff sample had the highest concentration observed (220 μg/L). Durum (1974) stated that concentrations of the carbonate and hydroxide forms of cadmium, with pH values equal to or less than 7, are relatively high, and that waters exceeding the U.S. Public Health Service (USPHS) standard of 10 μg/L may occur in many stable water systems, including both surface and groundwaters.

Chromium was detected by Pitt, et al. (1995) at concentrations above 1 μg/L in almost all of the stormwater source area sheetflow and receiving water samples an-

Table 14. New York City Combined Sewer Overflow Quality Summary (average of observed values)

	Total Sample		Filtered Sample[a]		
	Mean of Observ.	Detection Frequency	Mean of Observ.	Detection Frequency	Percent Filterable[b]
Microtox Toxicity					
(I_{35}, % light decrease)	59	20 of 20	—	—	—
pH	6.8	20 of 20	—	—	—
Suspended solids (mg/L)	94	20 of 20	—	—	—
Turbidity (NTU)	22	20 of 20	—	—	—
Particle size (median, microns)	50	20 of 20	—	—	—
Base Neutrals					
(1 µg/L detection limit)					
nitrobenzene	26.5	1 of 20	nd[c]	0 of 20	0
isophorone	10.4	1 of 20	nd	0 of 20	0
bis (2-chloroethyl) ether	15.5	1 of 20	nd	0 of 20	0
1,3-dichlorobenzene	22	1 of 20	nd	0 of 20	0
naphthalene	7.7	1 of 20	nd	0 of 20	0
diethyl phthalate	103	1 of 20	nd	0 of 20	0
fluorene	9.3	1 of 20	nd	0 of 20	0
di-n-butyl phthalate	33.2	4 of 20	nd	0 of 20	0
phenanthrene	33.2	1 of 20	nd	0 of 20	0
benzyl butyl phthalate	82.3	1 of 20	nd	0 of 20	0
fluoranthene	6.6	1 of 20	nd	0 of 20	0
bis (2-ethyl hexyl) phthalate	11500	5 of 20	nd	0 of 20	0
pyrene	15.3	1 of 20	nd	0 of 20	0
di-n-octyl phthalate	42.6	1 of 20	nd	0 of 20	0
benzo (a) anthracene	10.9	1 of 20	nd	0 of 20	0
chrysene	8.2	1 of 20	nd	0 of 20	0
Pesticides					
(0.3 µg/L detection limit)					
BHC	0.3	1 of 20	nd	0 of 20	0
DDD	1.2	1 of 20	nd	0 of 20	0
chlordane	0.5	1 of 20	nd	0 of 20	0
Heavy Metals					
(1 µg/L detection limit)					
aluminum					
(5 µg/L detection limit)	1890	20 of 20	204	14 of 20	11
cadmium					
(0.1 µg/L detection limit)	2.8	20 of 20	0.9	20 of 20	32
chromium	21.7	20 of 20	13.9	9 of 20	64
copper	93.5	20 of 20	12.8	20 of 20	14
lead	45.3	20 of 20	3.5	14 of 20	8
nickel	15.3	20 of 20	8.8	19 of 20	58
zinc	116	20 of 20	35.5	20 of 20	31

[a] A split sample portion was also filtered through a 0.45-µm membrane filter before analysis to determine the filterable pollutant concentrations.

[b] The "percent filterable" is the percentage of the total sample concentration associated with the filtered sample portion: (filtered sample conc./total sample conc.)×100.

[c] ND: not detected.

Source: Pitt and Barron 1990.

Table 15. Median Concentrations Observed at Toronto Outfalls During Warm Weather[a]

Constituent (mg/L unless noted)	Warm Weather Baseflow		Warm Weather Stormwater	
	Residential	Industrial	Residential	Industrial
stormwater volume (m³/ha/season)	—	—	950	1500
baseflow volume (m³/ha/season)	1700	2100	—	—
total residue	979	554	256	371
total dissolved solids (TDS)	973	454	230	208
suspended solids (SS)	<5	43	22	117
chlorides	281	78	34	17
total phosphorus	0.09	0.73	0.28	0.75
phosphates	<0.06	0.12	0.02	0.16
total Kjeldahl nitrogen (organic N plus NH₃)	0.9	2.4	2.5	2.0
ammonia nitrogen	<0.1	<0.1	<0.1	<0.1
chemical oxygen demand (COD)	22	108	55	106
fecal coliform bacteria (#/100 mL)	33,000	7,000	40,000	49,000
fecal strep. bacteria (#/100 mL)	2,300	8,800	20,000	39,000
pseudo. aeruginosa bacteria (#/100 mL)	2,900	2,380	2,700	11,000
arsenic	<0.03	<0.03	<0.03	<0.03
cadmium	<0.01	<0.01	<0.01	<0.01
chromium	<0.06	0.42	<0.06	0.32
copper	0.02	0.05	0.03	0.06
lead	<0.04	<0.04	<0.06	0.08
selenium	<0.03	<0.03	<0.03	<0.03
zinc	0.04	0.18	0.06	0.19
phenolics (μg/L)	<1.5	2.0	1.2	5.1
α - BHC (ng/L)	17	<1	1	3.5
γ - BHC (Lindane) (ng/L)	5	<2	<1	<1
chlordane (ng/L)	4	<2	<2	<2
dieldrin (ng/L)	4	<5	<2	<2
polychlorinated biphenols (PCBs) (ng/L)	<20	<20	<20	33
pentachlorophenol (PCP) (ng/L)	280	50	70	705

[a] Warm weather samples were obtained during the late spring, summer, and early fall months when the air temperatures were above freezing and no snow was present.

Source: Pitt and McLean 1986.

alyzed. Landscaped area samples had the highest median total chromium concentrations observed (100 μg/L), while a roof runoff sample had the highest sample concentration observed (510 μg/L). Phillips and Russo (1978) stated that in water, trivalent (Cr^{+3}) chromium exists as a complex, colloid, or precipitate, depending on pH. The more toxic hexavalent (Cr^{+6}) chromium form is usually present only as an ion and would therefore not be directly removable through filtering or sedimentation practices.

Copper was detected by Pitt, et al. (1995) at concentrations greater than 1 μg/L in practically all of the stormwater samples analyzed. Urban creek samples contained the greatest median concentrations (160 μg/L), while a street runoff sample had the highest copper concentration observed (1250 μg/L). Callahan, et al. (1979) stated that copper in unpolluted waters occurs mostly as a carbonate complex and in polluted waters forms

Table 16. Median Concentrations Observed at Toronto Outfalls During Cold Weather[a]

Constituent (mg/L unless noted)	Cold Weather Baseflow		Cold Weather Snowmelt	
	Residential	Industrial	Residential	Industrial
stormwater volume (m³/ha/season)	—	—	1800	830
baseflow volume (m³/ha/season)	1100	660	—	—
total residue	2230	1080	1580	1340
total dissolved solids (TDS)	2210	1020	1530	1240
suspended solids (SS)	21	50	30	95
chlorides	1080	470	660	620
total phosphorus	0.18	0.34	0.23	0.50
phosphates	<0.05	<0.02	<0.06	0.14
total Kjeldahl nitrogen (organic N plus NH₃)	1.4	2.0	1.7	2.5
ammonia nitrogen	<0.1	<0.1	0.2	0.4
chemical oxygen demand (COD)	48	68	40	94
fecal coliform bacteria (#/100 mL)	9800	400	2320	300
fecal strep bacteria (#/100 mL)	1400	2400	1900	2500
pseudo. aeruginosa bacteria (#/100 mL)	85	55	20	30
cadmium	<0.01	<0.01	<0.01	0.01
chromium	<0.01	0.24	<0.01	0.35
copper	0.02	0.04	0.04	0.07
lead	<0.06	<0.04	0.09	0.08
zinc	0.07	0.15	0.12	0.31
phenolics (mg/L)	2.0	7.3	2.5	15
α - BHC (ng/L)	NA[b]	3	4	5
γ - BHC (Lindane) (ng/L)	NA	NA	2	1
chlordane (ng/L)	NA	NA	11	2
dieldrin (ng/L)	NA	NA	2	NA
pentachlorophenol (PCP) (ng/L)	NA	NA	NA	40

[a] Cold weather samples were obtained during the winter months when the air temperatures were commonly below freezing. Snowmelt samples were obtained during snowmelt episodes and when rain fell on snow.

[b] NA: not analyzed.

Source: Pitt and McLean 1986.

complexes with organic materials. Pitt and Amy (1973) found that inorganic copper is mostly found with valence states of plus one and plus two in natural water systems near pH 7. The common inorganic copper forms at these pH values include copper combined with sulfides, sulfates, oxides, hydroxides, cyanides, and iodide. Phillips and Russo (1978) stated that alkalinity and pH are believed to be the major factors controlling copper speciation. Callahan, et al. (1979) stated that copper speciation with organics is most important in polluted waters. Cu^{+2} is especially likely to form complexes with humic materials.

The EPA (1976) stated that the ferrous form of iron can persist in waters void of dissolved oxygen, and originates usually from anaerobic groundwaters or from mine drainage. Iron can exist in natural organometallic, humic, and colloidal forms. Black or brown "swamp waters" may contain iron concentrations of several milligrams per liter in the presence or absence of dissolved oxygen, but this iron form has little effect on aquatic life because it is complexed and relatively inactive chemically and physiologically.

Table 17. Toronto Cold Weather Snowmelt Source Area Sheetflow Quality (median observed concentrations, mg/L)

Source Area	Total Solids	Total Dissolved Solids (TDS)	Suspended Solids (SS)	Chlorides	Total Phosphorus	Phosphates	TKN	Ammonia	COD
Industrial									
Pervious Areas:									
Grass/open areas	390	282	77	100	0.33	0.10	1.4	<0.1	47
Unpaved storage/parking	2925	1000	2105	113	1.1	0.46	5.3	0.2	160
Impervious Areas:									
Sidewalks	1050	200	847	48	0.45	0.20	1.6	<0.1	63
Paved pk./storage & driveways	1690	349	260	0.55	0.18	3.8	<0.1	135	
Road gutters	1320	575	625	230	0.60	0.15	1.8	<0.1	230
Residential/Commercial									
Pervious Areas:									
Grass/open areas	94	78	40	4.0	0.29	0.20	1.2	0.4	26
Impervious Areas:									
Sidewalks	390	29	281	6.4	0.63	0.38	2.6	2.6	98
Paved pk., driveways & loading	918	274	380	81	0.64	0.08	2.5	<0.1	110
Paved roads	890	166	284	56	0.30	0.06	1.8	<0.1	140
Road gutters	530	190	152	25	0.54	0.28	2.3	<0.1	66
Roadside grass swales	380	155	50	37	0.59	0.17	1.8	0.1	40

Table 17. Continued

Source Area	Fecal Coliforms	Fecal Strep.	Pseudo. aerug.	Cadmium	Chromium	Copper	Lead	Zinc	Phenolics (µg/L)
	(counts/100mL)								
Industrial									
Pervious Areas:									
Grass/open areas	<20	100	<20	<0.005	0.01	0.01	0.01	0.06	3.0
Unpaved storage/parking	<100	100	<20	0.011	0.07	0.13	0.26	0.51	9.0
Impervious Areas:									
Sidewalks	<50	<50	<20	<0.005	0.02	0.11	0.09	0.47	3.7
Paved pk./storage & driveways	<100	450	<20	<0.005	0.02	0.05	0.20	0.40	4.0
Road gutters	<100	100	<20	<0.005	0.05	0.12	0.45	0.66	9.0
Residential/Commercial									
Pervious Areas:									
Grass/open areas	<20	350	<10	<0.005	<0.01	<0.01	0.04	0.02	1.4
Impervious Areas:									
Sidewalks	75	600	<20	<0.005	<0.01	0.02	0.15	0.16	1.4
Paved pk., driveways & loading	<20	200	10	<0.005	0.02	0.04	0.23	0.23	2.6
Paved roads	50	200	<10	<0.005	0.01	0.05	0.26	0.26	3.2
Road gutters	60	4600	<10	<0.005	0.01	0.02	0.12	0.09	1.8
Roadside grass swales	60	1300	<10	<0.005	<0.01	0.01	0.05	0.08	1.6

Source: Pitt and McLean 1986.

Table 18. Toronto Warm Weather Source Area Sheetflow Quality (median observed concentrations, mg/L)

Source Area	Total Solids	Total Dissolved Solids (TDS)	Suspended Solids (SS)	Chlorides	Total Phosphorus	Phosphates	TKN	Ammonia	COD
Industrial									
Pervious Areas:									
Bare ground	488	240	247	400	0.62	0.20	2.7	0.2	40
Unpaved driveway & pk/storage	1148	420	805	1160	1.09	0.09	2.8	<0.1	247
Impervious Areas:									
Roofs	113	107	5	NA[a]	<0.05	<0.02	1.7	0.35	55
Sidewalks	580	145	435	257	0.82	0.03	4.7	<0.1	98
Paved pk./storage & driveways	315	112	202	240	0.9	0.06	3.1	0.15	132
Paved Roads	992	188	871	220	0.9	0.06	3.5	<0.1	326
Residential									
Pervious Areas:									
Bare ground	1240	436	807	250	0.20	0.66	1.3	0.5	66
Impervious Areas:									
Roofs	44	39	<3	56	<0.04	<0.02	0.8	0.1	36
Sidewalks	49	28	20	63	0.8	0.64	1.1	0.3	62
Paved driveways & parking	952	78	687	92	0.62	<0.02	2.2	<0.1	67
Paved roads	185	51	136	79	0.49	0.03	1.6	<0.1	66

Table 18. Continued

Source Area	Fecal Coliforms	Fecal Strep.	Pseudo. aerug.	Cadmium	Chromium	Copper	Lead	Zinc	Phenolics (µg/L)
	(1,000 counts/100 mL)								
Industrial									
Pervious Areas:									
Bare ground	3.3	43	2.1	<0.03	<0.15	<0.1	<0.3	0.05	0.8
Unpaved driveways & pk/storage	26	6.2	0.5	<0.004	<0.06	0.13	0.25	0.50	9.0
Impervious Areas:									
Roofs	1.6	0.6	0.06	<0.004	<0.06	0.015	<0.04	0.07	1.2
Sidewalks	55	3.6	3.7	<0.004	<0.06	0.03	0.04	0.06	8.7
Paved pk./storage & driveways	2.8	0.4	0.7	<0.004	<0.06	0.05	0.19	0.34	8.6
Paved roads	19	8.5	5.1	<0.004	<0.06	0.13	0.51	0.59	14.7
Residential/Commercial									
Pervious Areas:									
Bare ground	NA	NA	NA	<0.001	<0.01	0.02	0.03	0.04	<0.4
Impervious Areas:									
Roofs	0.5	0.9	0.1	<0.003	<0.03	0.01	<0.03	0.31	2.8
Sidewalks	11	1.8	0.6	<0.004	<0.06	0.02	0.08	0.06	8.6
Paved driveways & parking	2.0	1.0	0.4	<0.004	<0.06	0.05	0.35	0.45	11.8
Paved roads	4.8	7.9	0.1	<0.004	<0.03	0.02	0.13	0.16	6.3

[a] NA = not analyzed

Source: Pitt and McLean 1986.

Table 19. Ottawa Sheetflow Bacteria Characteristics (August 15 and September 23, 1981 samples)

	Rooftop Runoff	Vacant Land and Park Runoff	Parking Lot Runoff	Gutter Flow
Fecal Coliforms				
geometric mean (#/100 mL)	85	5,600	2,900	3,500
min (#/100 mL)	10	360	200	500
max (#/100 mL)	400	79,000	19,000	10,000
number of observations	4	7	6	7
Fecal Strep				
geometric mean (#/100 mL)	170	16,500	11,900	22,600
min (#/100 mL)	20	12,000	1,600	1,800
max (#/100 mL)	3,600	57,000	40,000	1,200,000
number of observations	4	7	6	7

Source: Pitt 1983.

Pitt, et al. (1995) found lead at concentrations greater than 1 μg/L in all of the stormwater runoff and CSO samples analyzed. Vehicle service area samples contained the greatest median concentrations (75 μg/L), while a storage area runoff sample had the highest lead concentration observed (330 μg/L). Lead exists in nature mainly as lead sulfide (galena) (EPA 1976). Other common natural forms of lead are lead carbonate (cerussite), lead sulfate (anglesite) and lead chlorophosphate (pyromorphite). Stable complexes result from the interaction of lead with organic materials. The toxicity of lead in water is affected by pH, hardness, organic materials, and the presence of other metals. Pitt and Amy (1973) reported that most inorganic lead in runoff water systems near pH 7 exists in the plus 2 or plus 4 valence states as lead sulfide, carbonate, sulfate, chromate, hydroxide, or chloride. Rolfe and Reinhold (1977) found that about 46% of the total lead input in a test watershed remained airborne. The total input included gaseous and particulate vehicle emissions. About 5% of the total lead input to the watershed occurred with rainfall and about 60% occurred with atmospheric settleable particulates. The streamflow accounted for the majority of all of the lead discharged from the watershed (about 7 to 8% of the total lead input).

Nickel was detected at concentrations greater than 1 μg/L in most of the stormwater samples analyzed by Pitt, et al. (1995). Parking area runoff samples contained the greatest median concentrations (40 μg/L), while a landscaped area runoff sample had the highest nickel concentration observed (130 μg/L).

Zinc was detected by Pitt, et al. (1995) at concentrations greater than 1 μg/L in all of the stormwater runoff samples analyzed. Roof runoff samples contained the greatest median concentration (100 μg/L) and the highest concentration observed (1580 μg/L). The EPA (1976) stated that zinc is usually found in nature as the sulfide. It is often associated with the sulfides of other metals, especially lead, copper, cadmium, and iron. Callahan, et al. (1979) stated that zinc in unpolluted waters is mostly as the hydrated divalent cation (Zn^{+2}) but in polluted waters complexation of zinc predominates. Pitt and Amy (1973) reported that zinc is mostly found as the divalent form, as a sulfide, oxide, sulfate, or hydroxide.

The polycyclic aromatic hydrocarbon (PAH) compounds found in urban runoff (most commonly anthracene, chrysene, fluoranthene and phenanthrene) are formed by in-

Table 20. Birmingham, AL, Source Area Sheetflow Quality (Average Concentrations of Observed Compounds)

Pollutant (µg/L unless noted)	Residential Roofs		Commercial Roofs		Industrial Roofs		Resid./Inst. Streets		Industrial Streets		Resid. Pvd. Parking		Com./Inst. Pvd. Parking	
	Total	Filtered	Total	Filtered	Total	Filtered	Total	Filtered	Total	Filtered	Total	Filtered	Total	Filtered
microtox toxicity														
(I_{35}, % light decrease)	41	—	29	—	20	—	9.5	—	34	—	37	—	30	—
pH (pH units)	6.5	—	6.2	—	8	—	7.2	—	7.9	—	7.1	—	6.8	—
suspended solids (mg/L)	26.8	—	3	—	5	—	14.5	—	65.5	—	15.5	—	41	—
turbidity (NTU)	4.2	—	3.3	—	4	—	5.5	—	48.1	—	15.3	—	7.9	—
particle size (median, microns)	22.4	—	27	—	14	—	30	—	27	—	35	—	34.4	—
Metals:														
aluminum	1910	362	152	82	319	167	181	155	4520	1250	2500	610	558	94.4
cadmium	6.4	0.14	0.43	0.08	1.2	0.47	0.46	0.31	55.7	0.25	35.3	0.2	2.6	0.5
chromium	12.2	0.86	170	0.85	6.2	0.72	3	0.9	11	1.6	290		19.6	1.5
copper	46	2.8	5.1		240	1.6	10	1.34	410	4.2	285	2.1	39.3	15.8
lead	51		28.9		37.8	0.66	15.8	2.73	56.3	0.9	66.7	1	45.4	1.5
nickel	15		24.4		4		2.23		19.9		35.3		33.1	2.8
zinc	476	438	181	128	33.5	21.8	37.5	37.5	67.5	27.3	64	55.5	178	144
Base-Neutrals:														
bis (2-chloroethyl) ether	4.5		29.2	5.9	5.5				4.2	1			2.4	
1,3-dichlorobenzene	14	3.8	29.8	7.9					1.6	1.1	21.4	5.1	8	2.5
bis (2-chloroisopropyl) ether	46		22.9								41		40.2	
bis (2-chloroethoxy) methane														
hexachloroethane			18.9		12.3	5.7							11.4	
naphthalene			62.7	4.6									9.4	1.3
di-n-butyl phthalate	6.6													
acenaphthylene														
fluorene														
phenanthrene			7.7								20.8		2	
anthracene			8.2											
benzyl butyl phthalate			35.2						0.5		1.9	1.9	3.1	
fluoranthene			15.3	1.9	6	3.8					47.4	2.7	2.4	

bis (2-ethyl hexyl) phthalate						153			
pyrene	5.5	9.8	40.3	0.6	0.6		9.5		
benzo (a) anthracene	2.5		28.3				5.8		
chrysene			14.9				24.7		
benzo (b) fluoranthene	3.8		66.4		3.9		89	7.3	4.4
benzo (k) fluoranthene	6.6		5.9		4.2		74.1	3.3	3.4
benzo (a) pyrene	5.5		39.4		5.1		100		9.3
benzo (g,h,i) perylene			10.1					13.3	
Pesticides:									
α-BHC							0.29		
γ-BHC							0.39		
aldrin							0.29		
DDT			0.23		0.31		0.66	0.2	9.4
chlordane								0.15	0.37
methoxychlor	0.36								
endrin	0.17	0.18	0.78						

Table 20. Continued

Pollutant (µg/L unless noted)	Inst. Unpvd. Parking		Indus. Unpvd. Parking		Com./Indus. Pvd. Storage		Indus. Unpvd. Storage		Loading Docks		Vehicle Service		Landscaped Areas	
	Total	Filtered	Total	Filtered	Total	Filtered	Total	Filtered	Total	Filtered	Total	Filtered	Total	Filtered
microtox toxicity (I_{35}, % light decrease)	29	–	18	–	45	–	46	–	30	–	25	–	24	–
pH (pH units)	8.2	–	7.9	–	9.1	–	8.2	–	7.8	–	7.2	–	6.6	–
suspended solids (mg/L)	391	–	170	–	15	–	152	–	40	–	23.8	–	37.6	–
turbidity (NTU)	392	–	42.4	–	10.2	–	81.7	–	16.5	–	10.2	–	55	–
particle size (median, µm)	38	–	42.7	–	47.3	–	23.8	–	31.7	–	37	–	29.2	–
Metals:														
aluminum	11600	370	3140	1070	514	267	2940	52.9	777	7.7	705	136	2310	1210
cadmium	1.5	1.5	1	1.5	4.6	1.2	6.7	2.2	1.4	0.44	9.2	0.23	0.29	0.24
chromium			6.5	1.9	33.8	3	100	9.4	17.1	5.9	74.3	0.9	94.4	1.8
copper	390	2.3	13.3	3.6	16.7	0.9	458	305	21.6	5.9	135	6.8	94.4	3.3
lead	65.5	0.9	28	1.7	21	1.7	155	2.8	55.1	1.1	63.4	1.3	28.5	0.7
nickel	30		73.3		30.6		69	17.8	6.7	0.8	42	6.6	38.2	
zinc	81.5	18	28.3	19	48	39.2	2740	7.8	36.8	22.2	105	72.8	263	165
Base-Neutrals:														
bis (2-chloroethyl) ether	12.2										9.4	4.9	11.7	
1,3-dichlorobenzene							3.6	3.3			28.8	10.6	13.5	2.6
bis (2-chloroisopropyl) ether											47.3		17.4	
bis (2-chloroethoxy) methane													3.8	1.6
hexachloroethane											11.9	11.1		
naphthalene											28.4	16.8	10.3	
di-n-butyl phthalate											0.8	0.8		
acenaphthylene											0.6			
fluorene											0.6			
phenanthrene											2.6		6	
anthracene											9.3	2.6	4.5	
benzyl butyl phthalate											10.5	4.2	26	

fluoranthene		1.3		15.9	1.7	8.2	0.7
bis (2-ethyl hexyl) phthalate	10.7						
pyrene		2		18.1	1.9	2.4	
benzo (a) anthracene				14.3		11.2	
chrysene				5.4			
benzo (b) fluoranthene				39.7		6.3	
benzo (k) fluoranthene				23.9		12.7	
benzo (a) pyrene				36.2		11.2	
benzo (g,h,i) perylene							
Pesticides:							
α-BHC							
γ-BHC							
aldrin							
DDT							
chlordane		0.89	0.43	0.28			
methoxychlor				0.18			
endrin							

Blanks: All samples had non-detectable quantities
    Heavy metals: 1 µg/L detection limits, except 5 µg/L for aluminum, and 0.1 µg/L for cadmium.
    Base neutrals: generally 1 µg/L detection limits.
    Pesticides: generally 0.3 µg/L detection limits.
Source: Pitt and Barron 1990.

complete combustion when organic compounds are burned with insufficient oxygen. Most of the PAHs are associated with suspended solids and humic materials, with little dissolved fractions found in natural waters. There are some studies that have examined the carcinogenic risk associated with the ingestion of PAHs by humans. Many animal studies have established the wide range of carcinogenicity of PAHs by skin contact and ingestion (Varanasi 1989). The concentrations of PAHs needed to produce cancers can be extremely low. As an example, the PAH concentration associated with a cancer risk level of 10^{-6} is only 9.7×10^{-4} µg/L. Tissue damage and systemic toxicity have also been associated with PAH exposure (PHS 1981).

Benzo (a) anthracene, a PAH, was detected at concentrations of about 2 to 60 µg/L in 12% of the stormwater samples analyzed by Pitt, et al. (1995). The greatest concentration observed was found in an urban creek sample. A major source of benzo (a) anthracene is gasoline, with an emission factor as high as 0.5 mg emitted in the exhaust condensate per liter of gasoline consumed (Verschueren 1983). Wood preservative use may also contribute benzo (a) anthracene. Typical domestic sewage effluent values ranged from 0.2 to more than 1 µg/L (in heavily industrialized areas). During heavy rains, sewage concentrations of benzo (a) anthracene increased substantially to more than 10 µg/L. Benzo (a) anthracene was reported to be both carcinogenic and mutagenic (Verschueren 1983).

Pitt, et al. (1995) detected benzo (b) fluoranthene in concentrations greater than about 1 µg/L in 17% of the stormwater samples analyzed. The greatest concentration observed was 226 µg/L, found in a roof runoff sample. Benzo (b) fluoranthene, a PAH, is found in gasoline, in addition to fresh and used motor oils (Verschueren 1983). The automobile emission factor for benzo (b) fluoranthene is about 20 to 50 µg in the exhaust condensate per liter of gasoline consumed. It is also found in bitumen, an ingredient of roofing compounds. Benzo (b) fluoranthene was found in domestic wastewater effluent in concentrations of about 0.04 to 0.2 µg/L. Raw sewage concentrations were as high as 0.9 µg/L in areas of heavy industry. Typical sewage concentrations were about 0.04 µg/L, but increased to about 10 µg/L during heavy rains. The IARC (1979) has found sufficient evidence of carcinogenicity of benzo (b) fluoranthene in animals.

Pitt, et al. (1995) detected benzo (k) fluoranthene in concentrations greater than 1 µg/L in 17% of all stormwater runoff samples analyzed. The greatest concentration observed was 221 µg/L, found in a roof runoff sample. Benzo (k) fluoranthene, a PAH, is found in crude oils, gasoline, and bitumen (Verschueren 1983). Sewage sludges have been found to contain from 100 to 400 µg/L benzo (k) fluoranthene. Domestic sewage effluent can contain from 0.03 to 0.2 µg/L benzo (k) fluoranthene, while sewage in heavily industrialized areas may contain concentrations as great as 0.5 µg/L. During heavy rains, sewage concentrations of benzo (k) fluoranthene increased to more than 4 µg/L.

Benzo (a) pyrene, a PAH, was detected by Pitt, et al. (1995) in concentrations greater than about 1 µg/L in 17% of the stormwater samples analyzed. The greatest concentration observed was 300 µg/L, found in a roof runoff sample. Benzo (a) pyrene can be synthesized by various bacteria (including *Escherichia coli*) at a rate of about 20 to 60 µg per dry kg of bacterial biomass (Verschueren 1983). It is also a potential leachate of asphalt and is present in oils and gasoline. Benzo (a) pyrene is present in domestic sewage effluents at concentrations of about 0.05 µg/L and in raw sewage sludge at concentrations of about 400 µg/L. Benzo (a) pyrene is a known carcinogen and mutagen.

Fluoranthene, a PAH, was detected in concentrations greater than about 1 µg/L in 23% of the stormwater samples analyzed by Pitt, et al. (1995). The greatest concentration observed was 128 µg/L, found in an urban creek sample. Fluoranthene was the only PAH found in an EPA drinking water survey of 110 samples in 1977 (Harris 1982). Fluoranthene is found in crude oils, gasoline, motor oils and wood preservatives (Verschueren 1983). It is found in the exhaust condensate of gasoline engines at a rate of about 1 mg per liter of gasoline consumed. It is found in domestic sewage effluents in concentrations of about 0.01 to 2.5 µg/L, and in raw sewage sludge at concentrations of up to about 1200 µg/L. In one case, sewage effluent had concentrations of fluoranthene of about 0.4 µg/L during dry weather, but increased to about 16 µg/L during heavy rains. Several studies have shown that fluoranthene is a potent carcinogen which substantially increases the carcinogenic potential of other known carcinogens (EPA 1980).

Naphthalene, a PAH, was detected in concentrations greater than about 1 µg/L in 13% of the runoff samples analyzed by Pitt, et al. (1995). The greatest concentration observed was 296 µg/L, found in an urban creek sample. Naphthalene is the single most abundant component of coal tar, and is present in gasoline and insecticides (especially mothballs). Naphthalene may also originate from natural uncontrolled combustion, such as forest fires, along with house fires in urban areas (Howard 1989). However, vehicle emissions are probably the most significant urban source of naphthalene. Additional major urban naphthalene sources included detergents, solvents, and asphalt (Verschueren 1983). Carcinogenicity and mutagenicity tests were negative for naphthalene (Howard 1989).

Phenanthrene, a PAH, was detected in concentrations greater than about 1 µg/L in 10% of the runoff samples analyzed by Pitt, et al. (1995). The greatest concentration observed was 69 µg/L, found in an urban creek sample. Phenanthrene is found in crude oil, gasoline, and coal tar. Its emission factor in gasoline engine exhaust condensate is about 2.5 mg per liter of gasoline consumed. Carcinogenicity and mutagenicity tests were negative for phenanthrene (Verschueren 1983).

Pyrene, a PAH, was detected in concentrations greater than about 1 µg/L in 19% of the runoff samples analyzed by Pitt, et al. (1995). The greatest concentration observed was 102 µg/L, found in an urban creek sample. Pyrene is found in crude oils, gasoline, motor oils, bitumen, coal tar, and wood preservatives (Verschueren 1983). The emission factor of pyrene from gasoline engine exhaust condensates is about 2.5 mg per liter of gasoline consumed. It was degraded in seawater by 85% from an initial concentration of 365 µg/L after 12 days. Pyrene is discharged in domestic wastewater effluents at concentrations of about 2 µg/L. In one study, dry-weather raw sewage had pyrene concentrations of about 0.2 µg/L, while pyrene concentrations in raw sewage during a heavy rain increased to about 16 µg/L. Mutagenicity test results of pyrene were negative, but pyrene is considered a human carcinogen (Verschueren 1983).

Pitt, et al. (1995) detected chlordane in concentrations greater than about 0.3 µg/L in 13% of the stormwater samples analyzed. The greatest concentration observed was 2.2 µg/L, found in a roof runoff sample. Chlordane is a nonsystemic insecticide and its registered use has been canceled by the EPA. The food chain concentration potential of chlordane is considered high. The EPA has also revoked chlordane residual tolerances in foods (Federal Register, Vol. 51, No. 247, page 46665, Dec. 24, 1986).

Butyl benzyl phthalate (BBP), a phthalate ester, was detected in concentrations greater than about 1 µg/L in 12% of the stormwater samples analyzed by Pitt, et al. (1995). The greatest concentration observed was 128 µg/L, found in a landscaped area runoff sample. BBP is used chiefly as a plasticizer in polyvinylchlorides (Verschueren 1983). BBP is not tightly bound to the plastic and is readily lost and enters aqueous solutions in contact with the plastic. The typical average concentration of BBP in natural U.S. waters is about 0.4 µg/L, but has been reported to be as high as 4.1 µg/L.

Bis (2-chloroethyl) ether (BCEE) was detected in concentrations greater than about 1 µg/L in 14% of the stormwater samples analyzed by Pitt, et al. (1995). The greatest concentration observed was 204 µg/L, found in an urban creek sample. BCEE is used as a fumigant, and as an ingredient in solvents, insecticides, paints, lacquers, and varnishes (Verschueren 1983). It is also formed by the chlorination of waters that contain ethers.

Bis (2-chloroisopropyl) ether (BCIE) was detected by Pitt, et al. (1995) in stormwater at concentrations greater than about 1 µg/L in 14% of the samples analyzed. The greatest concentration observed was 217 µg/L, found in a parking area runoff sample. BCIE was not found to be carcinogenic during rat tests (HEW 1979).

1,3-Dichlorobenzene (1,3-DCB) was detected in concentrations greater than about 1 µg/L in 23% of the stormwater samples analyzed by Pitt, et al. (1995). The greatest concentration observed was 120 µg/L, found in an urban creek sample.

Chapter 3

POTENTIAL GROUNDWATER CONTAMINATION ASSOCIATED WITH URBAN RUNOFF

This chapter addresses several categories of constituents that are known to affect groundwater quality, or the operation of infiltration or recharge devices, as documented in the groundwater contamination literature. The categories that can adversely affect groundwater quality include nutrients, pesticides, other organics, pathogens, metals, and salts and other dissolved minerals. Suspended solids, dissolved oxygen, and the sodium adsorption ratio (the ratio of monovalent, Na^+, to divalent cations, Ca^{+2} and Mg^{+2}) are also important for the operation of recharge and infiltration devices. The intention of this chapter is to identify known stormwater contaminants as to their potential to contaminate groundwater. Many of the references describe groundwater contamination problems with these pollutants from sanitary sewage and agricultural sources, not specifically from stormwater sources. Therefore, care must be taken when assuming that similar problems would occur with stormwater sources. Major differences between stormwater and these other sources which may affect groundwater contamination likely include the rate of pollutant application, intermittent versus continuous applications, and the presence of interfering compounds. However, the information included in this chapter enables the recognition of pollutants which should be considered when investigating stormwater infiltration. The National Research Council (1994) also presents additional information on the health risks associated with groundwater recharge using waters of impaired quality, including stormwater.

GROUNDWATER CONTAMINATION ASSOCIATED WITH NUTRIENTS

Definition

Primary nutrients are defined as "compounds or constituents that contain nitrogen (N), phosphorus (P) and other elements that are essential for plant growth" (Hampson 1986). Other needed aspects of nutrient use include the organic matter and bacteria that are needed to convert the primary nutrient from its natural form to a form that the organism can use. Nitrogen-containing compounds of interest are primarily from fertilizers and sanitary sewage, with the available nitrogen forms being nitrate, nitrite, ammonium, and urea. The phosphorus-containing compounds of interest are generally found in fertilizers, with the available phosphorus form being orthophosphate. Nitrogen and phosphorus are cyclic elements in that the combined forms may be changed and metabolized by decomposition and synthesis (Reichenbaugh 1977).

Examples of Nutrients Contaminating Groundwaters

Nutrients can originate from many different sources, including natural occurrence, sanitary sewage discharges and combined sewage overflows, landscaping/lawn maintenance and other urban sources (including septic tank and sewer system leakage, waste decomposition, and highway runoff) plus agricultural sources. Nitrates are one of the most frequently encountered contaminants in groundwater (AWWA 1990).

Phosphorus compounds of interest are associated with phosphorus-containing fertilizers (Lauer 1988a and 1988b) and detergents. Groundwater contamination by phosphorus has not been as widespread, or as severe, as that from nitrogen compounds. Nitrogen loadings are usually much greater than phosphorus loadings, especially from nonagricultural sources (Hampson 1986). Spray-irrigation with secondary-treated sanitary wastewater was found to increase both the total nitrogen and nitrate concentrations in a shallow aquifer in Florida, but these, and the total phosphorus concentrations, were significantly reduced within 200 feet of the test site (Brown 1982).

Natural Sources

Nitrogen occurs naturally both in the atmosphere and in the earth's soils. Natural nitrogen can lead to groundwater contamination by nitrates. As an example, in regions with relatively unweathered sedimentary deposits or loess beneath the root zone, residual exchangeable ammonium in the soil can be readily oxidized to nitrate if exposed to the correct conditions. Leaching of this naturally occurring nitrate caused groundwater contamination (with concentrations greater than 30 μg/L) in nonpopulated and nonagricultural areas of Montana and North Dakota (Power and Schepers 1989).

Forms of nitrogen from precipitation may be either nitrate or ammonium. Atmospheric nitrate results from combustion, with the highest ambient air concentrations being downwind of power plants, major industrial areas, and major automobile activity. Atmospheric ammonium results from volatilization of ammonia from soils, fertilizers, animal wastes, and vegetation (Power and Schepers 1989).

Urban Areas

Roadway runoff has been documented as the major source of groundwater nitrogen contamination in urban areas of Florida (Hampson 1986; Schiffer 1989; and German 1989). This occurs from both vehicular exhaust onto road surfaces and onto adjacent soils, and from roadside fertilization of landscaped areas. Roadway runoff also contains phosphorus from motor oil use and from other nutrient sources, such as bird droppings and animal remains (Schiffer 1989). Nitrate has leached from fertilizers and affected groundwaters under various turf grasses in urban areas, including at golf courses, parks and home lawns (Petrovic 1990; Ku and Simmons 1986; Robinson and Snyder 1991).

Leakage from sanitary sewers and septic tanks in urban areas can contribute significantly to nitrate-nitrogen contamination of the soil and groundwater (Power and Schepers 1989). Nitrate contamination of groundwater from sanitary sewage and sludge disposal has been documented in New York (Ku and Simmons 1986; Smith and Myott 1975), California (Schmidt and Sherman 1987), Narbonne, France (Razack, et al. 1988), Florida (Waller, et al. 1987) and Delaware (Ritter, et al. 1989).

Elevated groundwater nitrate concentrations have been found in the heavily industrialized areas of Birmingham, UK, due to industrial area stormwater infiltration (Lloyd, et al. 1988). The deep-well injection of organonitrile- and nitrate-containing industrial wastes in Florida has also increased the groundwater nitrate concentration in parts of the Floridan aquifer (Ehrlich, et al. 1979a and 1979b).

Agricultural Operations

In the United States, the areas with the greatest nitrate contamination of groundwater include heavily populated states with large dairy and poultry industries, or states having extensive agricultural irrigation. Extensively irrigated areas of the United States include the corn growing areas of Delaware, Pennsylvania, and Maryland; the vegetable growing areas of New York and the Northeast; the potato growing areas of New Jersey; the tobacco, soybean, and corn growing areas of Virginia, Delaware, and Maryland (Ritter, et al. 1989); the chicken, corn, and soybean production areas in New York (Ritter, et al. 1991); the western Corn Belt states (Power and Schepers 1989); and the citrus, potato, and grape vineyard areas in California (Schmidt and Sherman 1987). Table 21 groups the states according to the percentage of wells in the state that have groundwater nitrate concentrations greater than 3.0 mg/L.

Nitrogen leaching from agricultural fertilizers to irrigation return/drainage waters and eventually to underlying aquifers has been documented in California (Schmidt and Sherman 1987), Arizona, and New Mexico (Sabol, et al. 1987). Groundwater nitrate contamination has increased in these areas since the 1930s, when large-scale irrigation was instituted (Sabol, et al. 1987). Typical nitrate concentrations in shallow vadose zone

Table 21. Groundwater Nitrate Contamination in the United States

Contaminated Well Percentage Between 0.0% and 10.0%

Alabama (7.4%)	Alaska (5.2%)
Florida (4.3%)	Georgia (4.8%)
Hawaii (9.1%)	Louisiana (2.4%)
Massachusetts (5.5%)	Michigan (3.9%)
Mississippi (1.8%)	Missouri (8.7%)
Nevada (8.4%)	New Hampshire (4.3%)
North Carolina (5.9%)	North Dakota (9.0%)
Ohio (8.5%)	Oregon (6.6%)
South Carolina (4.1%)	Tennessee (5.5%)
Vermont (6.9%)	Virginia (3.9%)
West Virginia (5.5%)	

Contaminated Well Percentage Between 10.1% and 20.0%

Arkansas (12.4%)	Connecticut (16.7%)
Idaho (14.6%)	Illinois (14.0%)
Indiana (11.1%)	Iowa (18.4%)
Kentucky (17.2%)	Maine (14.2%)
Montana (11.5%)	New Jersey (11.4%)
New Mexico (12.7%)	South Dakota (14.9%)
Utah (10.4%)	Wisconsin (18.9%)
Wyoming (11.4%)	

Contaminated Well Percentage Between 20.1% and 30.0%

Colorado (22.9%)	Maryland (28.8%)
Minnesota (20.2%)	Texas (23.5%)
Washington (22.9%)	

Contaminated Well Percentage Between 30.1% and 40.0%

California (32.6%)	Delaware (34.6%)
Nebraska (32.7%)	Oklahoma (35.9%)
Pennsylvania (30.3%)	Puerto Rico (35.4%)

Contaminated Well Percentage Between 40.1% and 50.0%

Arizona (49.3%)	New York (40.3%)
Rhode Island (45.1%)	

Contaminated Well Percentage Greater than 50.0%

Kansas (54.2%)

Source: Power and Schepers 1989.

water beneath agricultural fields in Nebraska exceed 10 mg/L (Power and Schepers 1989; Spalding and Kitchen 1988). Irrigation is necessary to leach and reduce salt accumulation in plant root zones in many agricultural areas (Power and Schepers 1989). This leaching generally leads to increased groundwater contamination. Similar processes are likely to occur under irrigated urban landscaped areas, including private homes, parks, and golf courses.

Rates of nitrate concentration increases in groundwater in Nebraska have been reported to be from 0.4 to 1.0 mg/L/year (as nitrogen) in well-drained agricultural soils. In areas with fine-textured soils and thick vadose zone sediments, the increase is smaller, from 0.1 to 0.2 mg/L/year (Spalding and Kitchen 1988).

Leachate from waste decomposition can also contribute nitrogen-containing compounds to groundwater. The urine of grazing animals was the source of groundwater nitrate contamination in New Zealand (Close 1987). Grazing cattle return to the soil between 75 and 80% of the nitrogen, phosphorus, and potassium from their food (Reichenbaugh 1977). Land spreading of animal waste from large-scale, concentrated dairy and poultry industries in the Northeast U.S., the Great Lakes States (Power and Schepers 1989), and Maryland (Ritter, et al. 1989) caused the nitrate contamination of groundwater. Poorly managed feedlots can cause enhanced nitrate production from animal wastes which in turn leach through soil during rainfalls and enter the groundwater (Power and Schepers 1989).

Nutrient Leaching and Soil Removal Processes

Whenever nitrogen-containing compounds come into contact with soil, a potential for nitrate leaching into groundwater exists, especially in rapid-infiltration wastewater basins, stormwater infiltration devices, and in agricultural areas. Nitrate is highly soluble (>1 kg/L) and will stay in solution in the percolation water, after leaving the root zone, until it reaches the groundwater. Therefore, vadose zone sampling can be an effective tool in predicting nonpoint sources that may adversely affect groundwater (Spalding and Kitchen 1988).

Urban Areas

Nitrogen-containing compounds in urban stormwater runoff may be carried long distances before infiltration into soil and subsequent contamination of groundwater, affecting South Carolina's approach to golf course stormwater management (Robinson and Snyder 1991). The amount of nitrogen available for leaching is directly related to the impervious cover in the watershed (Butler 1987). Nitrogen infiltration is controlled by soil texture and the rate and timing of water application (either through irrigation or rainfall) (Petrovic 1990; and Boggess 1975). Landfills, especially those that predate the RCRA Subtitle D regulations, often produce significant nitrogen contamination in nearby groundwater, as demonstrated in Lee County, Florida (Boggess 1975). Studies in Broward County, Florida, found that nitrogen contamination problems can also occur in areas with older septic tanks and sanitary sewer systems (Waller, et al. 1987).

Nutrient leachates usually move vertically through the soil and dilute rapidly downgradient from their source. The primary factors affecting leachate movement are the layering of geologic materials, the hydraulic gradients, and the volume of the leachate discharge. Sandy soils show less rapid dilution of the contaminant (mixing of leachate with groundwater), compared to limestone (Waller, et al. 1987), silts, clays, and organic-rich sediments (Wilde 1994).

Once the leachate, or the waste liquid from an industrial injection well, is in the soil/groundwater system, decomposition by denitrification can occur, with the primary decomposition product being elemental nitrogen (Hickey and Vecchioli 1986). As an example, deep well injection of organonitriles and nitrates in a limestone aquifer acts like an anaerobic filter with nitrate-respiring bacteria being the dominant microorganism. These bacteria caused an 80% reduction of the waste within 100 m of injection in the Floridan aquifer, near Pensacola (Ehrlich, et al. 1979b).

Gold and Groffman (1993) reported groundwater leaching losses from residential lawns to be low for nitrates (typically <2 mg/L), when using application rates recommended for residential lawn care.

Agricultural Areas

Nitrogen entry, use, and removal in agricultural soils is best described in terms of the nitrogen cycle: plant uptake, atmospheric loss (NH_3 volatilization and denitrification), soil storage, runoff into surface water and/or leaching into groundwater (Petrovic 1990). Nitrogen leaching from soils is common in irrigated agricultural areas. Besides supplying required moisture, irrigation is also needed to prevent the accumulation of salts in the crops' root zone. If salt flushing occurs when there is nitrate in the root zone, the nitrates are leached farther into the soil and potentially into the groundwater (Power and Schepers 1989).

Irrigated areas in the Midwest and Upper Midwest States have a greater potential to leach nitrates to the groundwater than most other areas because: (1) irrigation is concentrated in areas with high soil hydraulic conductivities; (2) irrigated lands generally receive heavy applications of nitrogen fertilizers to increase crop yield to offset the high cost of irrigation; and (3) irrigation accelerates the movement of nitrates, other soluble constituents, and percolating water to the groundwater (Mossbarger and Yost 1989). These irrigated areas also have higher water tables than nonirrigated arid areas, further increasing the nitrate leachate potential.

The mass of nitrate leached to groundwater during irrigation is related to the drainage volume (Ritter, et al. 1989). Nitrates that are already in the soil or groundwater also travel faster during periods of maximum recharge (when the water table is highest) (Boggess 1975).

Factors that influence the degree of nitrogen leaching in agriculture areas are soil type, irrigation amounts and practices, nitrogen source and application rate, and the season of application. Significant leaching occurs during the cool, wet seasons. Cool temperatures reduce denitrification and ammonia volatilization, and limit microbial nitrogen immobilization and plant uptake; however, they also limit nitrification. The combination of low evapotranspiration and high precipitation means that more water drains out of the root zone into the vadose zone and groundwater (Petrovic 1990). Humid agricultural areas have shown a greater potential for nutrient leaching when compared to arid environments because of the availability of water for leaching during the cool seasons (Close 1987). Land use and soil permeability greatly affect groundwater chemistry, and ultimately its contamination potential, as shown in the Coastal Plains of the East Coast by Ritter, et al. (1989) and in California by Schmidt and Sherman (1987).

Although no one has indicated that the potential for nitrate contamination can be prevented, adjustments in fertilizer practices may control the rate of degradation of groundwater (Power and Schepers 1989). Use of slow-release fertilizers is recommended in areas with significant leaching problems, such as coastal golf courses. The slow-release fertilizers include urea formaldehyde (UF), methylene urea, isobutylidene diurea (IBDU), and sulfur-coated urea. The fast-release fertilizers that are not recommended include calcium nitrate, sodium nitrate, ammonium sulfate and urea (Horsley and Moser 1990).

Methods of fertilizer application also affect groundwater contamination. Side-dressing of fertilizers can increase groundwater nitrate concentrations faster than fertigation, the mixing of fertilizers with irrigation water (Ritter, et al. 1991). However, other researchers (Saffigna and Keeney 1977) found that side-dressing can reduce leaching, if irrigation is irregular. No difference in the depth of penetration of phosphorus was found between applying the same amount of fertilizer in one application versus the same amount spread over several applications (Lauer 1988a). Frequent fallowing of the soil (leaving it unplanted) was found to contribute to nitrate contamination below the root zone. Cultivation (plowing or disking), as part of the planting process, increases aeration and mixes crop residues (which are readily available carbon sources) with soil organisms. A flush of mineralization and nitrification usually occurs after cultivation and results in the accumulation of leachable nitrate in the soil (Power and Schepers 1989). Nitrate contamination, however, appears to be controlled more by denitrification than by preferential flow through the soil (Steenhuis, et al. 1988).

Controlled application of wastewater sludge is an effective fertilizer; it has not reduced crop yields where it has been applied and it does not contaminate the groundwater to any greater extent than chemical fertilizer applications. Sludge applications to crop lands are generally controlled by heavy metal accumulations in the soil and plants, not by groundwater contamination of nutrients. Like chemical fertilizers, sludge use in amounts that provide soil with nitrogen in excess of what plants need results in leaching. Liquid sludges mineralize nitrogen faster than composted sludges, with the application rate and soil type not affecting the mineralization rate (Chang, et al. 1988). The decomposition potential in the groundwater below wastewater irrigation areas is considerable, as the decomposition is controlled by bacteria that are already in the wastewater (Wolff 1988):

$$5 \; C + 4 \; NO_3 + 2 \; H_2O \rightarrow 2 \; N_2 + 4 \; HCO_3 + CO_2$$

Decomposition of crop waste affects nitrogen leaching potential. Nonlegume residues, such as cornstalks, decompose much more slowly and initially immobilize inorganic nitrogen in the biomass (Power and Schepers 1989). The ammonium cation (NH_4^+) tends to decompose more readily to form ammonia in alkaline soils than in acidic soils (White and Dornbush 1988).

General

During percolation through the soil, some nutrients are removed and the nutrient concentrations affecting the groundwater are significantly reduced. Phosphorus, in the form of soluble orthophosphate, may be either directly precipitated or chemically adsorbed onto soil surfaces through reactions with exposed iron, aluminum or calcium on solid soil surfaces (Crites 1985). Phosphorus fixation is a two-step process—sorption onto the soil solid and then conversion of the sorbed phosphorus into mineraloids or minerals. If the sorption sites are filled either with phosphate anions or other ions, phosphorus sorption will be low. The sorption of phosphorus per unit of percolation liquid decreases with each year of recharge (White and Dornbush 1988).

Downward movement of phosphorus in different soils was found to be directly related to the reactivity index measured for each soil, especially for surface-applied phosphorus fertilizer. In Washington, a difference in depth of penetration was noted, however, between sandy and clayey textured soils, with sandy textured soils showing the greater depth of penetration. When the fertilizer was surface applied, instead of sprinkler applied, and the soil was not inverted, most of the phosphorus remained within the top 5 to 7.5 cm of the surface (Lauer 1988b).

If the nitrogen is not used by the plant, it will leach through the soil toward the groundwater, with some being removed in the soil prior to its reaching the aquifer. Under certain conditions, losses of dissolved nitrate and nitrite could be described by zero-order kinetics (Hampson 1986). In general, however, the process is regulated by so many limiting factors that such a simplified description is not possible. Residual nitrate concentrations were found to be highly variable in soil due to factors such as soil texture, mineralization, rainfall, irrigation, organic matter content, crop yield, nitrogen fertilizer/sludge rate, denitrification, and soil compaction (Ferguson, et al. 1990). Nitrate's flow to groundwater from stormwater infiltration is controlled by the rate and volume of infiltration, horizontal and vertical groundwater flow, the depth to the water table, and the existence of areas/channels of preferential flow. Once the nitrate has reached the groundwater, its concentration may be reduced by dispersion, diffusion, and mixing with uncontaminated groundwater (Wilde 1994). However, on sludge application plots, it was noted that if the soil receives the same annual amount of organic residue, it will accumulate organic nitrogen until the equilibrium concentration is reached. As shown in Riverside, California, once the sludge application is terminated, the organic nitrogen will continue to mineralize for many years and crop yields will continue to be enhanced (Chang, et al. 1988). The amount of ammonia volatilization is influenced by the position of the nitrogen in the soil/turf grass after application. This position is highly influenced by rainfall and/or irrigation (Bowman, et al. 1987, as reported by Petrovic 1990).

Nitrate concentrations in the vadose zone of an agricultural field generally are highest near the surface, although during irrigation and in times of winter recharge, nitrate can be leached below the zone, where it will be taken up by crops or denitrified. In such cases, nitrate will eventually reach groundwater if excess water is added to overlying soil. As nitrogen passes through the soil, NH_4^+ is removed by cation exchange between the NH_4^+ and the H^+ on the soil (Ragone 1977). Nitrogen is also removed by the soil aquifer treatment process through denitrification, a biological process that needs anaerobic conditions and organic carbon for food for the denitrifying bacteria, which in turn produce free nitrogen gas and nitrous oxides (Bouwer 1985).

Phosphorus concentrations generally decrease with depth in agricultural soils because phosphorus is adsorbed to soil minerals and also precipitates readily with calcium, iron, or aluminum (Lauer 1988b and Ragone 1977). The dominant precipitation reactions are pH dependent, forming mostly iron and aluminum phosphates in acidic soils and calcium phosphates under alkaline conditions. In neutral soils, the precipitation reactions are strongly rate-limited, so that the apparent solubility of the phosphate compounds is higher than under either acid or alkaline conditions (Bouwer 1985).

Health Problems

Excessive nutrient concentrations can cause both environmental and human health problems.

Environmental Problems

Nitrogen and phosphorus are fertilizers in aquatic environments, just as they are on land. The continual fertilization of the aquatic environment increases production of less desirable species and alters the aquatic community structure (Nightingale and Bianchi 1977a). Excess agricultural nitrogen applications can reduce both yields and quality of cotton, tomatoes, sugar beets, sugar cane, potatoes, citrus, avocados, peaches, apricots, apples, and grapes (Bouwer 1987). Nitrate ingestion by livestock has been linked to health problems, although livestock can tolerate up to ten times the maximum permissible concentrations allowed for humans with no significant adverse effects (Mossbarger and Yost 1989).

Excess agricultural phosphorus may reduce crop yields because the excess phosphorus reduces the availability to the plants of some micronutrients, such as copper, iron, and zinc. Excess soil phosphorus also increases calcium precipitation (Bouwer 1987).

Human Health Problems

Excess nitrate concentrations in water (>10 mg/L as N) consumed by infants causes methemoglobinemia, or "blue-baby disease." In infants, nitrate is converted to nitrite, which binds to red blood cells at the oxygen binding sites and asphyxiates the child by causing insufficient oxygen adsorption (Crites 1985). Nitrates may also increase the risk of stomach cancer through nitrate reduction to nitrite in the mouth or stomach. The nitrites then react with the amines to form carcinogenic nitrosamines (Bouwer 1989).

GROUNDWATER CONTAMINATION ASSOCIATED WITH PESTICIDES

Definition

Pesticides are generally classified into one of the following three groups, depending on their targets: herbicides, fungicides, or insecticides. Table 22 lists pesticides of potential concern in groundwater contamination studies and their typical uses from Environment Canada/Agriculture Canada (1987) and Pierce and Wong (1988).

Examples of Pesticide Contamination of Groundwaters

Urban Areas

Pesticides are used in urban areas, primarily for weed and insect control in houses, along roadsides, in parks, on golf courses, and on private lawns (Racke and Leslie 1993). The pesticide loading in runoff water has been correlated to the amount of impervious cover and to the distance the runoff will travel prior to infiltration or decomposition, as demonstrated by Lager (1977) and confirmed in Austin, Texas by Butler (1987). Urban pesticide contamination of groundwater in central Florida likely resulted from munici-

Table 22. Pesticides of Potential Concern in Groundwater Contamination Studies

Herbicides	Major Commercial Use
alachlor	weed control in beans, corn, potatoes, and soybeans
atrazine	grassy weed control in corn; soil sterilant on non-crop land
bromoxynil	weed control in barley, canary seed, corn, flax, oats, and seed-producing grasses
2,4-D	weed control in field crops and on non-crop land; soil sterilant; aquatic weed control (restricted use)
difenzoquat	post-emergent wild oat control in barley, canary grass, and wheat
diclofop-methyl	annual grass control in alfalfa, barley, soybeans, vegetable, wheat, flax, and canola
glyphosate	nonselective weed control in field crops, non-crop land and turf
MCPA	weed control in alfalfa, barley, corn, flax, oats, rye, wheat, and pastures
metolachlor	weed control in corn, soybeans, and potatoes
triallate	pre-emergent wild oat control in barley, flax, mustard, peas, canola, sugar beets, and wheat
trifluralin	pre-emergent weed control in field crops, vegetables and ornamentals
Insecticides	**Major Commercial Use**
carbaryl	specific insect control in livestock buildings, field crops, and fruits and vegetables
carbofuran	specific root worm, maggot, beetle, and leaf hopper control in vegetables; grasshopper, alfalfa weevil, and other insect control in field crops (restricted product)
chlorpyrifos	specific insect control in field crops, vegetables, and fruits; seed treatment for corn, beans, and peas; mosquito control
diazinon	specific insect control on fruits, vegetables, turf, and non-crop land; insect control in livestock buildings; seed treatment
fenitrothion	specific insect control on fruits, vegetables, turf, and non-crop land; insect control in livestock buildings; seed treatment
fonofos	specific insect control in corn, onions, potatoes, and tobacco
lindane	insect control on livestock, lawns, certain grains; seed treatment
malathion	specific insect control on livestock and in certain field crops
phorate	specific insect control on beans, corn, lettuce, and potatoes
terbufos	specific worm control in corn and sugar beets; flea beetle control in mustard and canola
Fungicides	**Major Commercial Use**
captan	specific fungal disease control on potato seed pieces, flower, fruit, vegetables, turf, and tobacco
chlorothalonil	fungal disease control on vegetables, potatoes, tomatoes, turf, and conifers
mancozeb	specific fungal disease control on various fruits, vegetables, and corn and potato seeds
maneb	specific fungal disease control on certain fruits and vegetables; seed treatment for barley, flax, oats, rye, sugar beets, and wheat
metiram	specific fungal disease control on potato seed pieces; seed treatment for barley, flax, oats, and wheat
thiram	specific fungal disease control in turf; seed treatment for vegetables, mustard, canola, barley, wheat, rye, oats, flax, corn, soybeans, alfalfa, and fruits

Adapted from Environment Canada/Agriculture Canada (1987) and Pierce and Wong (1988).

pal and homeowner use of these chemicals for pest control and their subsequent collection in stormwater runoff. Samples from the upper part of the Floridan aquifer have contained detectable amounts of diazinon, malathion, 2,4-D, ethion, methyl trithion, silvex, and 2,4,5-T (German 1989). In California, chlordane groundwater contamination has been traced to its application adjacent to residential foundations where it had been used for termite and ant control (Greene 1992). Atrazine and simazine groundwater contamination was related to their use to control weeds along roadways (Domagalski and Dubrovsky 1992). In Arizona, diazinon, dacthal, and dioxathion were detected in stormwater runoff entering urban dry wells that recharge the aquifer (Wilson, et al. 1990). Diazinon (at 30 µg/L) and methyl parathion (at 10 µg/L) were detected in groundwater below municipal waste treatment plants in Florida that used land spreading or well injection of wastes (Pruitt, et al. 1985). Gold and Groffman (1993) reported groundwater leaching losses from residential lawns to be low for dicamba and 2,4-D (<1 µg/L), when using application rates recommended for residential lawn care.

In contrast, groundwater below Fresno, California, stormwater recharge basins contained only one of the organophosphorus pesticides—diazinon. None of the ten chlorinated pesticides (aldrin, chlordane, endosulfan I, endosulfan II, endosulfan sulfate, DDD-mixed isomers, DDT-mixed isomers, DDE-mixed isomers, gamma-BHC, and methoxychlor) and none of the chlorophenoxy herbicides were found (Nightingale 1987b; and Salo, et al. 1986).

Agricultural Operations

Groundwater contamination by agricultural use of pesticides has been documented in the United States and Canada. There have been numerous observations of pesticide contamination of groundwater, including: alachlor (Wisconsin), aldicarb (Wisconsin, Arkansas, and California), atrazine (Wisconsin), bromacil (South Carolina, Georgia, and Florida), DBCP (California and South Carolina), EDB (South Carolina, Georgia, and Florida), and metolachlor (Wisconsin) (Krawchuk and Webster 1987). San Joaquin Valley (California) groundwater has been contaminated by atrazine, bromacil, 2,4-DP, diazinon, DBCP, 1,2-dibromoethane, dicamba, 1,2-dichloropropane, diuron, prometon, prometryn, propazine, and simazine. Atrazine and simazine were detected in 37.5% of the shallow wells sampled (Domagalski and Dubrovsky 1992). In Bakersfield, California, EDB has been found in wells in concentrations of 4 µg/L or less (Schmidt and Sherman 1987). Sandy soils in the Coastal Plains States in the East and Southeast are also very susceptible to pesticide leaching, primarily because the aquifer is directly below the recharge areas (Ritter, et al. 1989). However, tile drain water samples in the southeastern desert valleys of California showed minimal pesticide residue (Schmidt and Sherman 1987).

The EPA has conducted extensive surveys to investigate the potential contamination of drinking water wells by agricultural pesticides (EPA 1990). Forty-six pesticides have been found in groundwater of 35 states. California had 31 pesticides detected, Illinois had 17, Minnesota and New York each had 14 pesticides detected, Wisconsin had 13 detected, and Iowa had 11 detected. Natarajan and Rajagopal (1993) summarized Iowa DNR studies that concluded that hydrology, pesticide chemistry, and time of sampling all affected the observed occurrence of agricultural pesticides in public drinking groundwater supplies. One or more pesticides were detected in less than 10% of the 865 public

Table 23. Mobility Class Definition

Class	K(d)	M(I)
I - Mobile	<0.1 to 1.0	0.1 to 1.0
II - Intermediate mobility	1.0 to 10.0	0.01 to 0.1
III - Low mobility	10.0 to 100.0	0.001 to 0.01
IV - Very low mobility	>100.0	<0.001

Where K(d) is the soil adsorption coefficient

M(I) is the mobility index (ratio of pollutant's migration velocity to migration velocity of water under saturated flow).

Source: Armstrong and Llena 1992.

drinking water supply wells tested by Iowa DNR (Iowa DNR 1988 and 1990). Atrazine occurred most often, and was found in concentrations as high as 21 µg/L. Almost all atrazine was found in concentrations at less than 1 µg/L. Infiltration of soluble pesticides through the soil column was found to be the major route of pesticides to Iowa's shallow groundwater, even though the entry of surface runoff through fractures was responsible for relatively high doses of some pesticides.

Pesticide Removal Processes in Soil

Heavy repetitive use of mobile pesticides, such as EDB, on irrigated and sandy soils likely contaminates groundwater. Fungicides and nematocides must be mobile in order to reach the target pest, and hence they generally have the highest contamination potential. Pesticide leaching depends on patterns of use, soil texture, total organic carbon content of the soil, pesticide persistence, and depth to the water table (Shirmohammadi and Knisel 1989). A pesticide leaches to groundwater when its residence time in the soil is less than the time required to remove it or transform it to an innocuous form by chemical or biological processes. The residence time is controlled by two factors: water applied and chemical adsorption to stationary solid surfaces. Volatilization losses of soil-applied pesticides can be a significant removal mechanism for compounds having large Henry's constants (K_h), such as DBCP or EPTC (Jury, et al. 1983). However, for mobile compounds having low K_h values, such as atrazine, metolachlor, or alachlor, it is a negligible loss pathway compared to the leaching mechanism (Alhajjar, et al. 1990).

Mobility of Pesticides

Estimates of pesticide mobility can be made based on the three removal mechanisms affecting organic compounds (volatilization, sorption, and solubility), as shown in Tables 23 through 25 (Armstrong and Llena 1992). Application methods and formulation state can also play a significant role in pesticide mobility. The type of pesticide formulation will affect their loss, with wettable powders exhibiting the greatest losses to the soil (Domagalski and Dubrovsky 1992). Residues of foliar-applied water-soluble pesticides appear in high concentrations in runoff (Pierce and Wong 1988). Pesticide movement can be retarded or enhanced depending upon the soil conditions (Alhajjar, et al. 1990).

Table 24. Organic Compound Mobility for Sandy Loam Soils

Mobile (Class I)

Organic Carbon = 0.01%

dicamba	cyanazine
dacthal	metolachlor
2,4-D	malathion
diazinon	atrazine
alachlor	acenaphthylene

Organic Carbon = 0.1%

dicamba	dacthal
2,4-D	diazinon
alachlor	atrazine
cyanazine	metolachlor

Organic Carbon = 1.0%

dicamba	2,4-D
diazinon	alachlor
dacthal	

Intermediate Mobility (Class II)

Organic Carbon = 0.01%

2,4'-dichlorobiphenyl	pentachlorophenol
fluorene	anthracene
phenanthrene	chlordane
2,4,4'-trichlorobiphenyl	fluoranthene
pyrene	benzo (a) anthracene
methoxychlor	bis (2-ethylhexyl) phthalate
2,3',4',5-tetrachlorobiphenyl	2,3,3',4',6-pentachlorobiphenyl

Organic Carbon = 0.1%

malathion	acenaphthylene
2-4'-dichlorobiphenyl	pentachlorophenol
fluorene	

Organic Carbon = 1.0%

atrazine
metolachlor
cyanazine

Low Mobility (Class III)

Organic Carbon = 0.01%

chrysene	2,2',4,4',5,5'-hexachlorobiphenyl
benzo (a) pyrene	benzo (k) fluoranthene
benzo (b) fluoranthene	benzo (ghi) perylene
2,2',3,4,4',5,5'-heptachlorobiphenyl	

Organic Carbon = 0.1%

anthracene	phenanthrene
chlordane	2,4,4'-trichlorobiphenyl
fluoranthene	pyrene
benzo (a) anthracene	methoxychlor
bis (2-ethylhexyl) phthalate	2,3',4',5-tetrachlorobiphenyl
2,3,3',4',6-pentachlorobiphenyl	

Organic Carbon = 1.0%

malathion	acenaphthylene
2,4'-dichlorobiphenyl	pentachlorophenol
fluorene	

Table 24. Continued

Very Low Mobility (Class IV)	
Organic Carbon 0.01%	
indeno (1,2,3-cd) pyrene	
Organic Carbon = 0.1%	
2,3,3',4',6-pentachlorobiphenyl	chrysene
benzo (a) pyrene	2,2',4,4',5,5'-hexachlorobiphenyl
benzo (b) fluoranthene	benzo (k) fluoranthene
2,2',3,4,4',5,5'-heptachlorobiphenyl	benzo (ghi) perylene
ideno (1,2,3-cd) pyrene	
Organic Carbon = 1.0%	
anthracene	phenanthrene
chlordane	2,4,4'-trichlorobiphenyl
fluoranthene	pyrene
benzo (a) anthracene	methoxychlor
bis (2-ethylhexyl) phthalate	2,3',4',5-tetrachlorobiphenyl
2,3,3',4',6-pentachlorobiphenyl	chrysene
benzo (a) pyrene	2,2',4,4',5,5'-hexachlorobiphenyl
benzo (b) fluoranthene	benzo (k) fluoranthene
2,2',3,4,4',5,5'-heptachlorobiphenyl	benzo (ghi) perylene
indeno (1,2,3-cd) pyrene	

Source: Armstrong and Llena 1992.

Leaching is enhanced in alluvial soils (Domagalski and Dubrovsky 1992), but if the vadose zone contains restricting layers, pesticide movement will be slower (Sabol, et al. 1987). Leaching is also enhanced by flood-irrigation, in areas needing high recharge rates, and in areas with preferential flow. The greatest pesticide mobility occurs in areas with coarse-grained or sandy soils without a hardpan layer, having low clay and organic matter content and high permeability (Domagalski and Dubrovsky 1992). Structural voids, which are generally found in the surface layer of finer-textured soils rich in clay, can transmit pesticides rapidly when the voids are filled with water and the adsorbing surfaces of the soil matrix are bypassed. This preferential (bypass) flow is demonstrated in areas where the observed mobility of the pesticide in the soil is greater than the predicted value. This flow occurs in structured, coarse-grained soils, or soils with cracks, root holes, or worm holes. It generally does not occur on continuously flooded loam soil. It likely results from unsaturated flow processes and is controlled by the mobile and immobile fractions, soil heterogeneity, and soil spatial variability. Preferential flow allows the pesticide to flow easily through the soil and bypass the area with the greatest microbial activity and degradation (Rice, et al. 1991; and Steenhuis, et al. 1988).

Pesticide transport past the root zone through the unsaturated zone depends on the lipophilic nature and other chemical characteristics of the compound, on how the compound is used in relation to climate and irrigation practices, and on the properties of the soil and aquifer media, including hydraulic conductivity and total organic carbon (Domagalski and Dubrovsky 1992). Basic (high pH) pesticides, such as atrazine, become more mobile in soils having high pH values. Acidic pesticides ionize, depending on their dissociation constants, to form cationic and anionic species, and under neutral soil conditions (pH of 5 to 9), the anionic form predominates. Since these anions are negatively charged, they do not adsorb onto the negatively charged clay mineral sur-

Table 25. Organic Compound Mobility Classes for Silt Loam Soils

Mobile (Class I)

Organic Carbon = 0.01%

dicamba	cyanazine
dacthal	metolachlor
2,4-D	malathion
diazinon	atrazine
alachlor	acenaphthylene
2,4'-dichlorobiphenyl	

Organic Carbon = 0.1%

dicamba	dacthal
2,4-D	diazinon
alachlor	atrazine
cyanazine	metolachlor

Organic Carbon = 1.0%

dicamba	2,4-D
diazinon	alachlor
dacthal	

Intermediate Mobility (Class II)

Organic Carbon = 0.01%

pentachlorophenol	methoxychlor
fluorene	anthracene
phenanthrene	chlordane
2,4,4'-trichlorobiphenyl	fluoranthene
pyrene	benzo (a) anthracene
bis (2-ethylhexyl) phthalate	2,3,3',4',6-pentachlorobiphenyl
2,3',4',5-tetrachlorobiphenyl	

Organic Carbon = 0.1%

malathion	acenaphthylene
2-4'-dichlorobiphenyl	pentachlorophenol
fluorene	anthracene
phenanthrene	

Organic Carbon = 1.0%

atrazine	cyanazine
metolachlor	malathion

Low Mobility (Class III)

Organic Carbon = 0.01%

chrysene	2,2',4,4',5,5'-hexachlorobiphenyl
benzo (a) pyrene	benzo (k) fluoranthene
benzo (b) fluoranthene	benzo (ghi) perylene
2,2',3,4,4',5,5'-heptachlorobiphenyl	indeno (1,2,3-cd) pyrene

Organic Carbon = 0.1%

chlordane	2,4,4'-trichlorobiphenyl
fluoranthene	pyrene
benzo (a) anthracene	methoxychlor
bis (2-ethylhexyl) phthalate	2,3',4',5-tetrachlorobiphenyl
2,3,3',4',6-pentachlorobiphenyl	2,2',4,4',5,5'-hexachlorobiphenyl

Organic Carbon = 1.0%

acenaphthylene	anthracene
2,4'-dichlorobiphenyl	pentachlorophenol
fluorene	phenanthrene

Table 25. Continued

Very Low Mobility (Class IV)	
Organic Carbon 0.1%	
chrysene	2,2',4,4',5,5'-hexachlorobiphenyl
benzo (a) pyrene	benzo (k) fluoranthene
benzo (b) fluoranthene	benzo (ghi) perylene
2,2',3,4,4',5,5'-heptachlorobiphenyl	indeno (1,2,3-cd) pyrene
Organic Carbon = 1.0%	
chlordane	2,4,4'-trichlorobiphenyl
fluoranthene	pyrene
benzo (a) anthracene	methoxychlor
bis (2-ethylhexyl) phthalate	2,3',4',5-tetrachlorobiphenyl
2,3,3',4',6-pentachlorobiphenyl	chrysene
benzo (a) pyrene	2,2',4,4',5,5'-hexachlorobiphenyl
benzo (b) fluoranthene	benzo (k) fluoranthene
2,2',3,4,4',5,5'-heptachlorobiphenyl	benzo (ghi) perylene
ideno (1,2,3-cd) pyrene	

Reference: Armstrong and Llena 1992.

faces. Similarly, acidic pesticides tend to be more mobile in neutral soils (Pierce and Wong 1988). Pesticide movement will often slow as the depth increases in the vadose zone, allowing more soil contact time. However, the soil still may not be able to completely remove the pesticide from the water due to reduced biodegradation activity deeper in the soil (Bouwer 1987).

Pesticide mobility comparisons have been performed for atrazine, metolachlor, and alachlor in the same soil type, and it was found that alachlor mobility > metolachlor >> atrazine (Alhajjar, et al. 1990), with faster movement generally occurring in sandy loam soils versus loam soils (Krawchuk and Webster 1987).

Restricted pesticide usage on coastal golf courses has been recommended by regulatory agencies. The slower moving pesticides were recommended, provided they were used in accordance with the label's instructions. These included the fungicides iprodione and triadimefon, the insecticides isofenphos and chlorpyrifos and the herbicide glyphosate. Others were recommended against, even when used in accordance with the label's instructions. These included the fungicides anilazine, benomyl, chlorothalonil and maneb and the herbicides dicamba and dacthal. No insecticides were on the "banned list" (Horsley and Moser 1990).

Solubility and Sorptivity of Pesticides

Leaching of the less water-soluble compounds is determined by the sorption ability of the chemicals to the soil particles, especially the colloids. The sorption ability of the pesticide determines whether it will remain in solution until it reaches the groundwater (Pierce and Wong 1988). Adsorption of a pesticide to the soil stops its travel with the percolating water and prevents its contamination of the groundwater (Bouwer 1987). In general, pesticides with low water solubilities, high octanol-water partitioning coefficients, and high carbon partitioning coefficients are less mobile. Also, in general, basic and nonionic water-soluble pesticides are lost in greater amounts in surface runoff than acidic and nonionic low to moderate water-soluble pesticides with less traveling through the soil toward the groundwater (Pierce and Wong 1988).

Adsorption and desorption control the movement of pesticides in groundwater (Sabatini and Austin 1988). Modeling of pesticide movement using physical non-equilibrium expressions for mass transfer and diffusion most closely mimics the actual movement in soil (Pierce and Wong 1988).

Decomposition of Pesticides

Pesticides decompose in soil and water, but the total decomposition time can range from days to years. Decomposition and dispersion rates in the soil depend upon many factors, including pH, temperature, light, humidity, air movement, compound volatility, soil type, persistence/half-life, and microbiological activity (Ku and Simmons 1986).

Historically, pesticides were thought to adsorb to the soil during recharge, with decomposition then occurring from the sorbed sites. The decomposition rates are a function of temperature, moisture, and organic content, with the microbiological community being stable. Decomposition half-lives of many pesticides have been determined. However, literature half-lives generally apply to surface soils and do not account for the reduced microbial activity found deep in the vadose zone (Bouwer 1987).

Pesticides with a 30-day half-life can show considerable leaching. An order of magnitude difference in half-life results in a five- to ten-fold difference in percolation loss (Knisel and Leonard 1989). Organophosphate pesticides are less persistent than organochlorine pesticides, but they also are not strongly adsorbed by the sediment and are likely to leach into the vadose zone, and possibly the groundwater (Norberg-King, et al. 1991).

As demonstrated in Central Florida and on Long Island, New York, sediment analysis in recharge basins shows sediment with significant organic content, indicating that basin storage and recharge may effectively remove a large percentage of the pesticides (Schiffer 1989; Ku and Simmons 1986). Most organophosphate and carbamate insecticides are regarded as nonpersistent, but they have been found in older, organic soils used for vegetable production and in the surrounding drainage systems (Norberg-King, et al. 1991). Studies of recharge basins in Nassau and Suffolk Counties on Long Island, New York, showed that the DDT concentration in each basin correlated well with the basin's age and that DDT can survive in recharge basins for many years (Seaburn and Aronson 1974). Residues of atrazine, triallate and trifluralin have carried over from year to year in Canadian field soils (Smith 1982, as reported by Pierce and Wong 1988). Carbofuran has also survived for long periods in soil in Manitoba, Canada, with resulting detection in the groundwater the following year (11.5 to 158.4 µg/L) (Krawchuk and Webster 1987).

Observed chlorothalonil in groundwater (10.1 to 272.6 µg/L) possibly resulted from one of three sources: (1) carryover in the soil from the previous year before leaching to the groundwater (most probable); (2) use in other fields and subsequent movement of the groundwater; and (3) movement of tile drain water through the soil to the area in question. The source water for removing the pesticide from the soil in Manitoba, Canada, the following year was believed to be snowmelt water that leached into the ground in the early spring (Krawchuk and Webster 1987).

The following pesticides used in the San Joaquin Valley, California, have high leaching potential: alachlor, aldicarb, atrazine, bromacil, carbaryl, carbofuran, carboxin, chlorothalonil, cyanazine, 2,4-D, dalapon, DCPA, diazinon, dicamba, 1,2-dichloro-propane, dinoseb, disulfoton, diuron, methomyl, metolachlor, metribuzin, oxamyl, simazine, tebruthiuron, and trifluralin (Domagalski and Dubrovsky 1992)—with some having already been detected in the groundwater. Controlling pesticide leaching to prevent future groundwater contamination requires a reevaluation of current agricultural and residential practices and implementation of the more progressive ones (Lee 1990).

Health Effects

Some pesticides affect only those workers directly working with them, while others affect people or animals near and/or downwind of the application site (Bouwer 1989). Pesticides have been linked to cancer, nervous system disorders, birth defects, and other systemic disorders. DBCP has been linked to male sterility in the manufacturer's employees and to cancer in animals (Sabol, et al. 1987). Aldicarb has not been found to be carcinogenic, but it is highly toxic and causes a reversible inhibition of cholinesterase, an enzyme necessary for nerve function (Bouwer 1987). Methyl parathion and carbofuran are the organic toxic agents in the Colusa Basin water in California. Studies of the aquatic toxicity of these compounds found their toxicological effects to be additive when used in mixtures (Norberg-King, et al. 1991). Table 26 lists the toxicological data for many common pesticides.

GROUNDWATER CONTAMINATION ASSOCIATED WITH OTHER ORGANIC COMPOUNDS

Definition

Organic compounds are defined as compounds that are comprised mainly of carbon, nitrogen, and hydrogen. The organic compounds discussed in this chapter are generally analyzed using GC/MSD techniques and are divided into several categories, depending on the analytical technique used or the chemical class. The most common organic compounds that have been investigated during groundwater contamination studies are listed in Table 27.

Examples of Organic Compounds Contaminating Groundwater

Many organic compounds are naturally occurring, although many of concern in groundwater contamination investigations are manmade. Sources of organic contaminants include natural sources, landfills, leaky sewerage systems, highway runoff, agricultural runoff, urban stormwater runoff, and other urban and industrial sources and practices.

Natural Occurrence

Organic compounds occur naturally from decomposing animal wastes, leaf litter, vegetation, and soil organisms (Reichenbaugh, et al. 1977). Aerosols in groundwater usually are from precipitation (Seaburn and Aronson 1974).

Table 26. Pesticide Toxicological Data

Pesticide	LD$_{50}$/Species
aldicarb	0.93 mg/kg rat
atrazine	3080 mg/kg rat
	200->5000 mg/kg quail/mallard
azinphos-methyl	11–20 mg/kg rat
bromoxynil	190 mg/kg rat
	>5000 mg/kg quail/mallard
bromoxynil Octanoate	260 mg/kg rat
carbaryl	850 mg/kg rat
	2290->10000 smg/kg quail/mallard
carbofuran	8.2–14.1 mg/kg rat
	0.397–46 mg/kg quail/mallard
chlorothalonil	>10000 mg/kg rat
chlorpyrifos	135–163 mg/kg rat
	15.9–492 mg/kg quail/mallard
2,4-D	375 mg/kg rat
	412->5000 mg/kg quail/mallard
decamethrin	128 mg/kg rat
diazinon	108 mg/kg rat
	3.54–101 mg/kg quail/mallard
dicamba	1040–2900 mg/kg rat
diclofop-methyl	557–580 mg/kg rat
difenzoquat	470 mg/kg rat
	>5000 mg/kg quail/mallard
disulfoton	2.6–12.5 mg/kg rat
EPTC	1630 mg/kg rat
fenitrothion	500 mg/kg rat
	652–1662 mg/kg quail/mallard
fonofos	8–17 mg/kg rat
	16.9–290 mg/kg quail/mallard
glyphosate	4320 mg/kg rat
	>5000 mg/kg quail/mallard
lindane	76–200 mg/kg rat
	490->2000 mg/kg quail/mallard
malathion	1200 mg/kg rat
	1485–2968 mg/kg quail/mallard
mancozeb	5000 mg/kg rat
MCPA	700 mg/kg rat
metiram	10000 mg/kg rat
metolachlor	2750 mg/kg rat
metribuzin	2200 mg/kg rat
paraquat	150 mg/kg rat
phorate	2 mg/kg rat
	0.62–575 mg/kg quail/mallard
propanil	1400 mg/kg rat
terbufos	4.5–9.0 mg/kg rat
	225 mg/kg quail/mallard
triallate	1675–2166 mg/kg rat
trifluralin	>10000 mg/kg rat

Source: Krawchuk and Webster 1987; Pierce and Wong 1988.

Table 27. Organic Compounds Investigated During Groundwater Contamination Studies

Volatile Organic Compounds

benzene	bromoform
carbon tetrachloride	chlorobenzene
chlorodibromomethane	chloroethane
2-chloroethylvinyl ether	chloroform
1,2-trans-dichloroethylene	dichlorobomomethane
dichlorofluoromethane	1,1-dichloroethane
1,2-dichloroethane	1,1-dichloroethylene
1,2-dichloropropane	1,3-dichloropropene
ethylbenzene	methyl bromide
1,1,2,2-tetrachloroethane	tetrachloroethane
toluene	1,1,1-trichloroethane
1,1,2-trichloroethane	trichloroethylene
trichlorofluoromethane	vinyl chloride
dibromochloropropane	

Acid-Extractable Organic Compounds

p-chloro-m-cresol	2-chlorophenol
2,4-dichlorophenol	2,4-dimethylphenol
4,6-dinitro-o-cresol	2,4-dinitrophenol
2-nitrophenol	4-nitrophenol
pentachlorophenol	phenol
2,4,6-trichlorophenol	2-ethylphenol
2-methylphenol	2,6-dimethylphenol
3-methylphenol	2,5-dimethylphenol
3,4-dimethylphenol	2,3-dimethylphenol
3,5-dimethylphenol	2,4,6-trimethylphenol
2,3,5-trimethylphenol	2,3,6-trimethylphenol
2,3,5,6-tetramethylphenol	2-naphthol

Base-Neutral Extractable Compounds

acenaphthene	acenaphthylene
antracene	benzidene
benzo (a) anthracene	benzo (a) pyrene
benzo (g,h,i) perylene	benzo (k) fluoranthene
benzo (b) fluoranthene	4-bromophenyl phenyl ether
butyl benzyl phthalate	bis (2-chloroethyoxy) methane
bis (2-chloroethyl) ether	bis (2-chloroisopropyl) ether
2-chloronaphthalene	4-chlorophenyl phenyl ether
chrysene	dibenzo (a,h) anthracene
di-n-butyl phthalate	1,3-dichlorobenzene
1,4-dichlorobenzene	1,2-dichlorobenzene
3,3'-dichlorobenzidene	diethyl phthalate
dimethyl phthalate	2,6-dinitrotoluene
2,4-dinitrotoluene	dioctylphthalate
bis (2-ethylhexyl) phthalate	fluoranthene
hexachlorobenzene	hexachlorobutadiene
hexachlorocyclopentadiene	hexachloroethane
indeno (1,2,3-cd) pyrene	isophrone
naphthalene	nitrobenzene
n-nitrosodi-n-propylamine	n-nitrosodimethylamine
n-nitrosodiphenylamine	phenanthrene
pyrene	1,2,4-trichlorobenzene
benzo (b) pyrene	butyl benzylphthalate

Source: German 1989; Troutman, et al. 1984; Wanielista, et al. 1991; and Salo, et al. 1986.

Urban Areas

Concentrations of organic compounds in urban runoff are related to land use, geographic location and traffic volume (Hampson 1986). These compounds result from gasoline and oil drippings, tire residuals and vehicular exhaust material (Seaburn and Aronson 1974; Hampson 1986). The primary source is from the use of petroleum products, such as lubrication oils, fuels, and combustion emissions (Schiffer 1989). The organic compounds on many street surfaces consist of: cellulose, tannins, lignins, grease and oil, automobile exhaust hydrocarbons, carbohydrates, and animal droppings (Hampson 1986). Toluene and 2,4-dimethyl phenol are also found in urban runoff and are used in making asphalt (German 1992). Polynuclear aromatic hydrocarbons (PAHs) are also commonly found in urban runoff and result from combustion processes, and include fluoranthene, pyrene, anthracene, and chrysene (German 1989; Greene 1992).

In Florida, organic compounds found in runoff were attenuated in the soil, with only one priority pollutant being detected in the Floridan aquifer as a result of stormwater runoff. This compound was bis (2-ethylhexyl) phthalate, which is a plasticizer which readily leaches from plastics (German 1989). In Pima County, Arizona, base-neutral compounds appeared in groundwater from residential areas, while phenols in the groundwater were noted only near a commercial site. Groundwater from a commercial site, also in Pima County, has been contaminated with ethylbenzene and toluene. Perched groundwater samples from residential sites showed the presence of toluene, xylene, and phenol (Wilson, et al. 1990). On Long Island, New York, benzene (groundwater concentrations of 2 to 3 µg/L); bis (2-ethylhexyl) phthalate (5 to 13 µg/L); chloroform (2 to 3 µg/L); methylene chloride (stormwater concentration of 230 µg/L and groundwater concentrations of 6 to 20 µg/L); toluene (groundwater concentrations of 3 to 5 µg/L); 1,1,1-trichloroethane (2 to 23 µg/L); p-chloro-m-cresol (79 µg/L); 2,4-dimethylphenol (96 µg/L); and 4-nitrophenol (58 µg/L) were detected in groundwater beneath stormwater recharge basins (Ku and Simmons 1986).

Organic compounds occasionally found in runoff at three stormwater infiltration sites in Maryland included benzene, trichlorofluoromethane, 1,2-dichloroethane, 1,2-dibromoethylene, toluene, and methylene blue active substances (MBAS). Only MBASs were found consistently and in elevated concentrations beneath the infiltration devices. The other organic compounds found in runoff were removed either in the device or in the vadose zone. Although specific organic compounds were not detected in the groundwater beneath and downgradient of the infiltration device, the dissolved organic carbon (DOC) concentration in the groundwater affected by infiltration was greater than that in the native groundwater (Wilde 1994).

Leaky sanitary sewerage in the Munich, Germany urban area has caused elevated concentrations in groundwater of total organic carbon, chloroform, trichloroethylene, and tetrachloroethylene (Merkel, et al. 1988). Volatile organic compounds (chloroform concentrations of 4.5 to 29 µg/L) were detected in the groundwater below 12 of 15 municipal wastewater treatment plants throughout Florida (Pruitt, et al. 1985).

Organic compounds can leach from municipal waste landfills and other disposal sites, including unlined industrial surface impoundments and older hazardous waste landfills. Municipal waste landfills are sources of phthalate compounds that leach from plastic and detergents. Bis (2-ethylhexyl) phthalate can become airborne during municipal waste incineration, can be leached from plastics in a landfill, and was detected in water samples from the Floridan aquifer (German 1989). Surface impoundments that were used to contain industrial wastewater at a wood treatment plant showed significant amounts of residual pentachlorophenol (PCP), creosote, and diesel fluids and has led to the phenolic contamination of the groundwater near a wood treatment facility near Pensacola, Florida (Troutman, et al. 1984).

Industrial areas contribute heavily to the organic compound load that could potentially leach to the groundwater. Surface impoundments may be used to contain industrial wastes, deep well injection may be used to dispose of water, and stormwater runoff may collect organics as it passes over an industrial site. Phenols and the PAHs benzo (a) anthracene, chrysene, anthracene and benzo (b) fluoroanthenene, have been found in groundwater near an industrial site in Pima County, Arizona. The phenols are primarily used as disinfectants and as wood preservatives and were present in the stormwater runoff, although they were significantly reduced in concentration by the time they reached the groundwater (generally less than 50 µg/L). At an Arizona recharge site, the groundwater has higher concentrations of trichloroethylene, tetrachloroethylene, and pentachloroanisole, than the inflow water, indicating past industrial contamination (Bouwer et al. 1984).

Documented industrial groundwater contamination is not limited to the United States. In Birmingham, UK, groundwater contamination resulted from hydrocarbon oil and volatile chlorinated solvent use. The metals-related industries have contributed significant amounts of trichloroethylene (groundwater concentrations of up to 4.9 mg/L have been noted) to the groundwater in this area, and since trichloroethylene has been replaced by 1,1,1-trichloroethane in industry, 1,1,1-trichloroethane contamination is beginning to occur. The other organic compound to show up in significant concentrations in Birmingham is perchloroethylene, a solvent used primarily in the dry cleaning laundry industry (Lloyd, et al. 1988). On the left bank of the Danube, the petrochemical refinery Slovnaft has contributed to groundwater contamination by leaking oil during tanker loading and unloading (Marton and Mohler 1988).

Soil Removal Processes for Organic Compounds

The appearance in groundwater of organic compounds, along with elevated nitrate concentrations, has been used as an indicator of groundwater contamination (Lloyd 1988). Most organics are reduced in concentration during percolation through the soil, although they may still be detectable in the groundwater. Groundwater contamination from organics, like from other pollutants, occurs readily in areas with pervious soils, such as sand and gravel, and where the water table is near the land surface (Troutman, et al. 1984). Based on septic tank effluent studies, sand seems to be more effective than limestone in filtering the organic material (Schneider, et al. 1987). In coastal areas and valleys, direct interaction of groundwater and surface water will result in groundwater contamination if the surface water is contaminated (Troutman, et al. 1984). Organic removal from the soil and recharge water can occur by one of three methods: volatilization, sorption, and/or degradation (Crites 1985; and Nellor, et al. 1985). Estimates of organic compound mobility can be made based on volatilization, sorption, and solubility, as previously shown in Tables 23 through 25 (Armstrong and Llena 1992).

Volatilization of Organics

The rate of volatilization is controlled by the compound's physical and chemical properties, its concentration, the soil's sorptive characteristics, the soil-water content, air movement, temperature, and the soil's diffusion ability. Volatilization can occur both during application and from soil sites after infiltration. Volatilization during application is controlled by the compound's physical and chemical characteristics, atmospheric conditions, and application method (Crites 1985) and has been measured by observing the reduction in the organic concentration across an infiltration basin.

Volatilization from sorbed sites of soils is a function of: transfer of the organic compound from the soil's sorbed sites to the solution, movement from the solution to the air trapped in the soil, and diffusion of the compound in the soil air to the atmosphere. The extent of each of the above reactions depends on the compound's solubility, its concentration gradient in the soil, and proximity of the molecule of interest to the soil surface (Crites 1985).

Volatile organic compounds are rarely found in stormwater recharge basins (<2.4 μg/L), dry wells (<175 μg/L), or the vadose zone or groundwater below the basins (<4 μg/L), as indicated by studies in Fresno, California, by Nightingale (1987b) and in Pima County, Arizona by Wilson, et al. (1990).

Sorption of Organic Compounds

There are at least six different sorption mechanisms: cation exchange, anion exchange, cation-dipole and coordination bonds, hydrogen bonding, van der Waals attraction, and hydrophobic bonding (Crites 1985). Sorption is a function of the following soil-water-compound system characteristics: sorbate shape/configuration (including structure and position of functional groups and presence and degree of molecular saturation); sorbate chemical characteristics (including acidity or basicity; water solubility; charge distribution; polarity and polarizability); and sorbent nature (including mineralogical composition, organic matter content and cation exchange capacity) (Crites 1985), with the clay and particulate organic matter content controlling the sorption (Bouwer et al. 1984).

Hydrophobic sorption onto organic matter was found to limit the mobility of less soluble base-neutral and acid extractable compounds through organic soils and the vadose zone in Orlando, Florida (German 1989). The degree of removal in soil of nonhalogenated organic compounds is greater than that of the halogenated organics (Bouwer, et al. 1984). Benzene, toluene and xylene were found in the soils and in the perched water table in Arizona and, as these compounds are relatively soluble, they may percolate easily through the vadose zone. The toluene concentration in the perched water table was 54 μg/L, while the toluene concentration in the water table was 3.7 μg/L (Wilson, et al. 1990).

Sorption is not always a permanent removal mechanism. A study in Florida has shown that organic solubilization can occur for several storms following dry periods. However, extended periods of complete aeration of bottom sediments may be counterproductive when trying to reduce organic compound concentrations (Hampson 1986).

Degradation and Decomposition of Organic Compounds

The third process for organic compound attenuation is chemical or biological degradation. Examples of chemical degradation processes include hydrolysis and photo-degradation. However, most of the trace organic removal is the result of biological degradation (Smith and Myott 1975). Many organics can be degraded by microorganisms, at least partially, but others cannot. Temperature, pH, moisture content, cation exchange capacity, and air availability may limit the microbial degradation potential for even the most degradable organics (Crites 1985). The end products of complete aerobic degradation include carbon dioxide, sulfate, nitrate, phosphate, and water, while the end products under anaerobic conditions include carbon dioxide, nitrogen, hydrogen sulfide, and methane (EPA 1992).

Conditions in the thick, aerobic, unsaturated zone provide a good environment for wastewater detergent concentration reduction through biochemical degradation and adsorption (Smith and Myott 1975). Halogenated one- and two-carbon aliphatic compounds are biotransformed under methanogenic, but not aerobic, conditions (Bouwer, et al. 1984). The rate of breakdown of chlorinated hydrocarbons in the soil increases with temperature, water content, and organic matter content (Bouwer 1987). Nonhalogenated hydrocarbons decreased 50 to 90% during percolation through the soil, with concentrations in the renovated water being detectable, but near or below the detection limit. The halogenated organic compounds generally decreased to a lesser extent during percolation.

The chlorinated aromatics are relatively refractory and mobile in the ground and have lesser concentration decreases than nonchlorinated aromatic hydrocarbons. Significant reductions of TOC concentrations occurred during the first several meters of soil percolation and a gradual decrease in TOC concentration occurred with longer underground travel times at the Phoenix 23rd Avenue recharge project site (Bouwer, et al. 1984).

Pretreatment of the recharge water in Arizona by chlorination resulted in higher chloroform concentrations and in the formation of three brominated trihalomethanes (Bouwer, et al. 1984). Ponds and lakes affected by stormwater runoff have a high potential for the formation of trihalomethanes (THMs) with chlorination because of the precursor organics existing in the stormwater. Factors which affect THM formation include the chlorine contact time, pH, and temperature. Precursors to THM formation in stormwater include algae, bacteria, and humic substances (Wanielista, et al. 1991).

At an injection well site in Florida, organic carbon compounds were microbially converted to carbon dioxide and ammonia within 100 meters of the injection well (Hickey and Vecchioli 1986). In general, deep well injection in Florida showed that organic compounds are being mineralized in the Floridan aquifer (Ehrlich, et al. 1979b).

Anaerobic methanogenic bacteria in a surface impoundment located at a wood treatment plant near Pensacola, Florida, reduced the total phenol concentration by 45%. However, the presence of pentachlorophenol inhibits methanogenesis, reducing the removal of some organics in the impoundment (Troutman, et al. 1984). In Arizona, partial degradation of the chlorinated benzenes occurs during percolation through the aerobic zone, but the poor overall removal efficiency of chlorinated aromatics probably results from their lack of degradation under anoxic conditions. In general, infiltration and percolation through the soil have the effect of dampening concentration fluctuations and eliminating occasional extreme values (Bouwer, et al. 1984).

Health Effects

Some of the organics listed as hazardous to human health include: benzene, ethylenimine, ethylene dibromide, benzidene, carbon tetrachloride, tricresyl phosphate, chloroform, allyl chloride, aroclor 1254, and benzolalpyrene (Crites 1985). Disease outbreaks due to water contamination by organics has been documented for the following compounds: cutting oil, photographic developer fluid (hydroquinone, paramethylamino phenol), ethyl acrylate, fuel oil, leaded gasoline, mixtures of lubricating oil and kerosene, phenol, and polychlorinated biphenyl. In Wisconsin, an accidental phenol spill caused an illness in the area residents that was characterized by diarrhea, mouth sores, burning of the mouth, and dark urine (Craun 1979). Runoff from paved areas in urban Arizona that is directed to dry wells commonly was found to contain the following four suspected carcinogenic PAHs: benzo (a) anthracene, benzo (b) fluoranthene, dibenzo (a,h) anthracene and chrysene (Wilson, et al. 1990).

GROUNDWATER CONTAMINATION ASSOCIATED WITH PATHOGENS

Definition

Microorganism pathogens that can be in stormwater and can cause specific human diseases include viruses, protozoa, and bacteria. When evaluating the potential contamination of groundwater, many types of microorganisms are important. Pathogenic microorganisms harmful to human health need to be controlled, while those that are necessary for decomposition of nutrients and contaminants need to be encouraged. Viruses of concern in water are shown in Table 28.

Microorganisms that have been studied in urban runoff (both the harmful and the environmentally necessary ones) include: *Shigella, Salmonella, Giardia lamblia, Yersinia enterocolitica, Escherichia coli,* methanogenic bacteria, *Azomonas, Acinetobacter, Pseudomonas, Bacillus, Flavobacterium, Alcaligenes, Thiobacillus, Leptothrix, Thiothrix, Siderocapsaceae, Metallogenium, Vibrio, Campylobacter, Leptospira, Streptococcus, Beggiatoa, Micrococcus, Proteus, Aeromonas, Serratia,* and *Gallionella.*

Examples of Microorganism Contamination of Groundwater

Pathogen sources that potentially contaminate groundwater include waste decomposition, sanitary sewage, agriculture, urban runoff, and natural occurrence.

Natural Occurrence

Many types of bacteria occur naturally in water. The bacteria groups found in flowing water include sheathed bacteria, iron-manganese bacteria and some of the sulfur-oxidizing bacteria (Wanielista, et al. 1991). Bacteria that affix to carbon and are water and soil chemoorganotrophs include *Azomonas, Acinetobacter, Pseudomonas, Bacillus, Flavobacterium* and *Alcaligenes.*

Table 28. Viruses of Concern in Water

Virus Group	Number of Types
Enteroviruses	
Poliovirus	3
Echovirus	34
Coxsackie Virus A	24
Coxsackie Virus B	6
New Enteroviruses	?
Hepatitis Type A (probably an enterovirus)	1
Rotavirus (reovirus family) (gastroenteritis Type B)	2
Reovirus	3
Adenovirus	>30
Parvovirus	3
Adeno-Associated Virus	?

Source: Crites 1985.

Beggiatoa are filamentous bacteria that produce polysaccharidic sheath-like slimes. They are found at the sulfide/oxic interface of fresh or marine waters or estuaries. *Gallionella* is chemolithotrophic (it oxidizes inorganic ferrous iron and assimilates carbon dioxide). It is characterized by twisted stalks containing the organic material to which the ferric hydrate is bound. *Leptothrix*, if it is the dominant bacteria in water, indicates a high iron and/or manganese concentration in the water. *Metallogenium* are aerobic, chemoorganotroph/parasitic organisms that are found in surface waters near anaerobic sediments, decomposing layers of leaf litter, or bottom deposits of lakes. The *Siderocapsaceae* family of bacteria are nonfilamentous, aerobic/anaerobic, facultative/obligate slime-encapsulated bacteria that precipitate iron and manganese. They occur in swamp ditches, stagnant waters, and in the hypolimnion of lakes, and use organic carbon of iron-manganese humates. *Thiobacillus* is an aerobic, nonaggregating bacteria which oxidizes reduced sulfur into sulfate and is found in sulfur-bearing water. *Thiothrix* bacteria are found in sulfide- or thiosulfide-containing waters and in wastewater treatment plants (Reichenbaugh, et al. 1977).

The bacteria found in the aquifer at a recharge site at Bay Park, New York, included obligate aerobes and facultative anaerobes such as *Alcaligenes, Micrococcus, Flavobacterium, Acinetobacter, Proteus, Aeromonas*, and *Serratia*. These bacteria used the soluble organic matter in the recharge water as food. Denitrifying strains of *Pseudomonas fluorescens* grow under the anaerobic conditions of the Magothy aquifer by using nitrate as a terminal electron acceptor. Several *Pseudomonas* species, including *P. fluorescens, P. aeruginosa, P. putida, P. alcaligenes,* and *P. pseudoalcaligenes* can sustain themselves under anaerobic conditions by decomposing arginine, an amino acid, through substrate level phosphoration (Ehrlich, et al. 1979a).

Urban Areas

When comparing urban runoff from different land uses on Long Island, New York, low-density residential and nonresidential areas contributed the fewest bacteria to the storm runoff, while medium-density residential and commercial areas contributed the most. The amount of each bacteria species in the runoff varied by season, with the warm seasons having significantly more fecal coliforms and fecal streptococci than the colder seasons. However, total coliform concentrations were not affected by the season (Ku and Simmons 1986). NURP monitoring results also found similar seasonal variations for fecal coliform discharges throughout the U.S. (EPA 1983).

Viruses were detected in groundwater on Long Island at sites where stormwater recharge basins were located less than 35 ft above the aquifer (Vaughn, et al. 1978). At other locations, viruses are likely removed from the percolation water by either adsorption and/or inactivation.

Rainwater infiltration through solid wastes readily carries decomposition products, in addition to bacteria and viruses, downward through the soil to the groundwater. The highest contaminant concentrations occur when the water table is near the land surface (Boggess 1975). Bacterial pathogens found in human and animal feces and the resulting wastewater include *Escherichia coli, Salmonella, Shigella, Yersinia, Vibrio, Campylobacter* and *Leptospira* (Amoros, et al. 1989). Leaky sanitary sewerage systems or septic tanks may leach these bacteria into storm drainage systems or directly into groundwater (Pitt, et al. 1993).

Shigella is not as common in wastewater as *Salmonella* in developed countries, although it is more prevalent in the tropics and subtropics. *Salmonella* is the most common disease-causing bacteria found in wastewater (Crites 1985).

Agricultural Operations

Agricultural uses of treated effluent in fertilization can also promote the spread of pathogens to the soil, where, during irrigation, they can pass through the soil to the groundwater (Robinson and Snyder 1991). Bacteria also can be returned to humans through vegetables irrigated with inadequately disinfected sewage effluent (Amoros, et al. 1989).

Soil Removal Processes for Pathogens

During land application or burial of sewage sludge, pathogens may leach from the sludge and contaminate groundwater, although most are removed in the soil during percolation (Gerba and Haas 1988).

The factors that affect the survival of enteric bacteria and viruses in the soil include pH, antagonism from soil microflora, moisture content, temperature, sunlight, and organic matter. The effect of each of these factors is given in Table 29 (Crites 1985). In general, drying of the soil will kill both bacteria and viruses.

Viruses

Viral adsorption—Viral adsorption is promoted by increasing cation concentration, decreasing pH and decreasing soluble organics (EPA 1992), and is controlled by both the efficiency of short-term virus retention and the long-term behavior of viruses in the soil (Crites 1985). The effect of each of these factors is given in Table 30 (EPA 1992 and Crites 1985).

Table 29. Soil Characteristics Affecting Pathogen Removal

Factor	Pathogen Type	Remarks
pH	bacteria	shorter survival in acid soils (pH = 3 to 5) than in alkaline and neutral soils
antagonism from soil microflora	viruses bacteria	insufficient data increased survival time in sterile soil
moisture content	viruses bacteria and viruses	insufficient data longer survival in moist soils and during periods of high rainfall
temperature	bacteria and viruses	longer survival at low (winter) temperatures
sunlight	bacteria and viruses	shorter survival at the soil surface
organic matter	bacteria and viruses	longer survival (regrowth of some bacteria) when sufficient amounts of organic matter are present

Source: Crites 1985.

Table 30. Factors That Influence Viral Movement in Soil and to Groundwater

Soil Type. Fine-textured soils retain viruses more effectively than light-textured soils. Iron oxides increase the adsorptive capacity of soils. Muck soils are poor adsorbents. The higher the clay content of the soil, the greater the expected removal of the virus. Sandy loam soils and other soils containing organic matter are also favorable for virus removal. Soils with a small surface area do not achieve good virus removal.

pH. Generally, adsorption increases when pH decreases. However, the reported trends are not clear due to complicating factors.

Cations. Adsorption increases in the presence of cations. (Cations help reduce repulsive forces on both virus and soil particles.) Rainwater may desorb viruses from soil due to its low conductivity.

Soluble organics. Organics generally compete with viruses for adsorption sites. No significant competition at concentrations found in wastewater effluents. Humic and fulvic acids reduce virus adsorption to soils.

Virus type. Adsorption to soils varies with virus type and strain. Viruses may have different isoelectric points.

Flow rate. High flow rates reduce virus adsorption to soils.

Saturated vs. unsaturated flow. Virus movement is less under unsaturated flow conditions.

Rainfall. Viruses retained near the soil surface may be eluted after a heavy rainfall because of the establishment of ionic gradients within the soil column.

Sources: EPA 1992; and Crites 1985.

The downward movement and distribution of viruses are controlled by convection and hydraulic dispersion mechanisms. Since the movement of viruses through soil to groundwater occurs in the liquid phase and involves water movement and associated suspended virus particles, the distribution of viruses between the adsorbed and liquid phases determines the viral mass available for movement.

The distribution (or sorption) of virus particles in the soil matrix is largely due to electrostatic double layer interactions and van der Waals forces. Adsorption of viruses in soil is rapid and reversible and can be adequately described by an equilibrium (linear Freundlich isotherm) expression (Tim and Mostaghim 1991). Viral adsorption by the soil does not necessarily result in virus inactivation and has been shown to be reversible after a change in soil conditions, such as the ionic environment (Jansons, et al. 1989b; and Vaughn, et al. 1978). Also, once the virus reaches the groundwater, it can travel laterally through the aquifer until it is either adsorbed or inactivated (Vaughn, et al. 1978).

Viral inactivation—Enterovirus survival in groundwater is highly variable and is influenced by a number of factors, including: virus type, pH, temperature, dissolved oxygen concentration, and microbial antagonism (Jansons, et al. 1989b; Tim and Mostaghimi 1991). The two most important attributes of viruses that permit their long-term survival in the environment are their structure and very small size. These characteristics permit virus occlusion and protection within colloid-size particles. Viruses in wastewater applied to the soil are integral parts of submicron particles and are small enough to move with the applied water to the groundwater (Wellings 1988).

Dissolved oxygen is a significant factor in loss of virus activity in groundwater, possibly because direct oxidation of components of the virus capsid inactivates the virus, or, as with temperature, the level of dissolved oxygen influences the activity of antagonistic microorganisms. At temperatures below 4°C, microorganisms can survive for months or even years, whereas at higher temperatures, inactivation or dieoff occurs rapidly. Decreasing pH promotes virus adsorption and results in shorter survival times, both of the virus and of the antagonistic bacteria. High levels of organic matter appear to shield viruses from adsorption (Treweek 1985). Virus inactivation in the subsurface environment can be described by a first-order decay reaction (Tim and Mostaghim 1991). It is difficult to describe a soil that will remove all viruses effectively, as different soil types affect each virus differently (Jansons, et al. 1989a).

Enteric viruses are more resistant to environmental factors than enteric bacteria, and they exhibit longer survival times in natural waters. They can occur in potable and marine waters in the absence of fecal coliforms. Enteroviruses are more resistant to commonly used disinfectants than are indicator bacteria, and can occur in groundwater in the absence of indicator bacteria (Marzouk, et al. 1979).

Removal of Bacterial Pathogens

The major bacterial removal mechanisms in soil are straining at the soil surface and at intergrain contacts, sedimentation, sorption by soil particles, and inactivation (Crites 1985). Potable water use requires continuous disinfection to prevent disease outbreaks in the population served (Craun 1979).

Not all bacteria are harmful and need to be removed from the water. For example, methanogenic bacteria are responsible for the decomposition of phenolic compounds in infiltrating water (Troutman, et al. 1984).

Bacterial movement—The characteristics of bacterial movement through porous media, such as an aquifer, include the following (Ehrlich, et al. 1979a):

- Bacteria travel with the flow of water, not against the hydraulic gradient.
- The rate of bacterial movement through filtering is a function of the nature of the aquifer materials, with silt and soils having high clay content being the best materials for bacterial removal.

- The rate of bacterial removal by filtering can be characterized by the filter efficiency of the aquifer (filterability).
- Typically, the maximum distance that bacteria travel in a porous medium ranges from 50 to 100 ft.
- Under favorable condition, bacteria may survive as long as five years.

Stable populations of bacteria in soil and water may become established due to an increased concentration of adsorbed organics and due to the large surface area of the granules to which they attach themselves (Wanielista, et al. 1991).

Bacteria survival—Factors such as temperature, pH, metal concentration, nutrient availability, and other environmental characteristics affect the ability of a bacterial colony to survive in the water or soil (Ku and Simmons 1986). Once the microorganisms are retained in the soil, their survival depends upon the sunlight exposure, oxidation rates, desiccation, and antagonism from the established soil microbial population (Crites 1985).

Bacteria survive longer in acidic soils and when large amounts of organic matter are present. Bacteria and larger organisms in wastewater are usually removed during percolation through a short distance of soil (EPA 1992). When recharging using deep well injection, the logarithm of the coliform bacteria density decreases linearly with distance from the recharge well (Ehrlich 1979).

In general, enteric bacteria survive in soil between two and three months, although survival times up to five years have been documented (Crites 1985). *E. coli* can survive and multiply on trapped organic matter, and *Salmonella* and *Shigella* have survived for 44 days and 24 days, respectively, at a recharge site in Israel (Goldschmid 1974).

Significant dieoff can be achieved using intermittent rapid infiltration of wastewater and allowing the recharge basin to dry out (Crites 1985). When the recharge is stopped, the bacteria multiply, using the accumulated organic debris for food. The decomposition of the organic matter starts as an aerobic process. After the oxygen has been consumed, the process turns anaerobic and septic. At this stage, some denitrification and sulfate reduction takes place and the decomposition of the organic matter will lead to a partial reduction in clogging. Pathogenic enterobacteria are unable to multiply under these conditions, but they can survive for a relatively long time. About a hundred days after recharge, the decomposition process ended and no more coliform bacteria were found in the pumped water from the recharged aquifer in Israel (Goldschmid 1974).

Health Effects

Human health can be affected through waterborne diseases transmitted by bacteria, viruses, protozoa, and parasites (Crites 1985). Many microorganism-caused diseases may be spread by person-to-person contact, or through contact of infected material (Gerba and Goyal 1988). Health effect research now goes beyond the acute effect of pathogens to the chronic problems associated with carcinogens and mutagens, and the other problems caused by long-term ingestion of low concentrations of these organisms (DeBoer 1983).

The development of clinical illness depends on numerous factors, including the immune status of the host, the age of the host, virulence of the microorganisms, type and strain of the microorganisms, and the route of infection. Mortality rates are affected by many of the same factors that determine the development of clinical illness (Gerba and Haas 1988).

Table 31. Enteroviruses and Diseases

Virus Group	Disease or Symptom
Poliovirus	paralysis, meningitis, fever, encephalitis, gastroenteritis
Echovirus	meningitis, respiratory diseases, rash, diarrhea, fever
Coxsackie Virus A	herpangina, respiratory disease, fever meningitis, myocarditis, diabetes
Coxsackie Virus B	myocarditis, congenital heart anomalies, rash, fever, meningitis, respiratory disease, pleurodynia, encephalitis
New Enteroviruses	meningitis, encephalitis, respiratory disease, acute hemorrhagic conjunctivitis, fever, gastroenteritis
Hepatitis Type A	infectious hepatitis
Rotavirus	epidemic vomiting and diarrhea, chiefly in children, gastroenteritis
Reovirus	respiratory infections
Adenovirus	respiratory disease, eye infections, conjunctivitis, gastroenteritis
Parvovirus	associated with respiratory diseases in children, but etiology not clearly established
Adeno-associated	associated with respiratory diseases in children, but etiology not clearly virus established
Norwalk virus	vomiting, diarrhea, fever, acute gastrointestinal disease
Astrovirus	gastroenteritis
Renovirus	not clearly established

Sources: Crites 1985; Tim and Mostaghim 1991.

Viruses

Enteric viruses can survive typical sewage treatment processes, including chlorination, in sufficient numbers to be detectable in the discharged water. Peak concentrations of these viruses occur in the late summer and early autumn. The documented outbreaks of viral diseases from groundwater contamination have been limited primarily to infectious hepatitis (Crites 1985; Craun 1979).

Human enteric viruses of concern, and their diseases or symptoms, are given in Table 31.

Bacteria

Shigella and *Shiga's bacillus* are dysentery bacilli and cause mild to acute diarrhea, abdominal pains, and blood in the stool. Acute gastroenteritis and typhoid fever result from the *Salmonella* bacteria. Symptoms of *Salmonella* contamination include abdominal cramps, nausea, vomiting, and diarrhea (Craun 1979). Most outbreaks of salmonellosis can be traced to transmission of the bacteria via food, milk, or direct contact, but its contamination of water supplies, with a resulting disease outbreak, occurred as recently as 1963. The bacteria *Escherichia coli* are responsible for some of the diarrhea diseases, as are *Shigella* and *Salmonella* (Crites 1985). *Azomonas, Acinetobacter, Pseudomonas bacillus* and *Flavobacterium* do not pose an immediate threat to human health, but they are opportunistic pathogens, chlorine resistant, and/or suppressers of total coliforms (Wanielista, et al. 1991).

Reviews of past disease outbreaks due to water contamination have shown that acute gastrointestinal illnesses were slightly higher in groundwater systems, compared to surface water systems, due to the lack of treatment, or the use of poor treatment practices. For surface water systems, contamination and the resulting outbreaks were caused by contaminated ice or containers, backsiphonage, cross-connections, water main breaks, or accidental contamination of treated storage reservoirs (Craun 1979).

GROUNDWATER CONTAMINATION ASSOCIATED WITH METALS

Even in small concentrations, metals may be a problem when infiltrating stormwater, especially when using a rapid infiltration system (Crites 1985). The heavy metals of most concern include lead, copper, nickel, chromium, and zinc. However, most of these metals have very low solubilities at the pH found in most natural waters, and they are readily removed by either sedimentation or sorption removal processes (Hampson 1986). Many are also filtered, or otherwise sorbed, in the surface layers of soils in infiltrating devices using surface infiltration.

Examples of Metals Contaminating Groundwater

Urban Areas

Nickel, chromium, and zinc concentrations exceeded the regulatory limits in the soil below a recharge area at an Arizona commercial site. However, only manganese was present at an elevated concentration in the groundwater at a residential site (Wilson, et al. 1990).

At a site in Lee County, Florida, groundwater near an unlined landfill had elevated iron concentrations due to landfill leachate. The leachate also may have increased the groundwater manganese concentrations (Boggess 1975). In New York, cesspool leachates have elevated the concentrations of boron and barium in the shallow groundwaters of the Magothy Aquifer (Smith and Myott 1975).

Boron concentrations were found to be high in groundwaters below industrialized areas in Birmingham, UK, as it is used in many metal-related industries. High concentrations of aluminum, cadmium, manganese, and titanium were also noted in the groundwaters near metals industries in the Birmingham industrial area (Lloyd, et al. 1988).

Agricultural Operations

In the Tulare Lake region of the Central Valley of California, the metals that have adversely affected groundwater quality include: boron, cadmium, chromium, copper, molybdenum, nickel, and selenium. Areas being irrigated had lower groundwater selenium concentrations, but had elevated concentrations of barium, molybdenum, vanadium, and zinc compared to nonirrigated areas (Deason 1989). At other agricultural sites, elevated groundwater concentrations of selenium, molybdenum, chromium, and mercury are of concern (Deason 1987).

Soils below drainage irrigation canals and basins have shown higher concentrations of selenium, arsenic, and uranium than normal for the Western United States (Deason 1989). Selenium groundwater contamination beneath irrigated lands was also documented in Wyoming (Peterson 1988).

Metal Removal Processes in Soils

The interaction of surface water and groundwater has resulted in selenium contamination of groundwater in Wyoming (Peterson 1988). Sandy soils resulted in minimal removal of boron and nickel, while the percolate water had no cobalt (Crites 1985). In general, studies of recharge basins receiving large metal loads show that most of the heavy metals are removed either in the basin sediment or in the vadose zone (Ku and Simmons 1986; and Hampson 1986).

Removal of metals by soil may be accomplished through one of several processes, including: (1) soil surface association, (2) precipitation, (3) occlusion with other precipitates, (4) solid-state diffusion into soil minerals, (5) biologic system or residue incorporation, and (6) complexation and chelation (Crites 1985). Most of these removal processes are pH-dependent, as is the solubility of most metals. In general, the solubility of a metal increases as the solution's pH decreases (Wilde 1994).

Adsorption

Dissolved metal ions are removed from stormwater mostly by adsorption onto the near-surface particles in the vadose zone, while the particulate metals are filtered out at the soil surface (Ku and Simmons 1986). Studies of dissolved lead ions in recharge ponds in Jacksonville, Florida, found that allowing the ponds to go dry between storms was counterproductive to the removal of lead from the water during recharge (Hampson 1986). Apparently, the adsorption bonds were weakened during the drying period. Studies in Fresno, California, recharge basins found that lead, zinc, cadmium, and copper accumulated at the soil surface with little downward movement over five years. However, the microtopographic features, such as small depressions and basin inlet and outlet locations, influenced the metal's distribution in the soil (Nightingale 1987a).

Similarities in water quality between the runoff water and the groundwater show that there is downward movement of copper and iron in sandy and loamy soils. However, the other metals of concern (arsenic, nickel, and lead) did not significantly move downward through the soil to the groundwater. The exception to this was some downward movement of lead with the percolation water in the sandy soils of Fresno stormwater recharge basins (Nightingale 1987b).

The order of sorption ability (from strongest to weakest) for potential substrates is as follows:

manganese oxides > organic matter > iron oxides > clay minerals

The order of sorption affinity (from strongest to weakest) for certain metals is as follows:

lead > copper > nickel > cobalt > zinc > cadmium > iron > manganese

Competition between substrates or between metals will affect the overall adsorption ability of various trace metals (Wilde 1994).

Cation Exchange, Organic Complexation, and Chelation of Metals

In soils, heavy metals enter into general cation exchange reactions with clay and organic matter and into chelation reactions with organic molecules. As the organic molecules are decomposed, the metals become free to react with iron and aluminum hydroxides, calcium, and other compounds. These new compounds are immobilized in the soil profile. The immobilization reactions are more pronounced at high pH and in an aerobic environment. Boron is adsorbed to iron and aluminum hydroxide coatings on clay minerals, to iron and aluminum oxides, to micaceous clay minerals, and to magnesium hydroxide coatings on weathering surfaces of ferro-magnesium minerals. In sandy soils and quartz, boron is not significantly immobilized (Bouwer 1985). Interactions of certain metals with phosphorus can form either soluble or insoluble complexes. Generally, the type of clay mineral also affects heavy metal adsorption, and the higher the cation exchange capacity (CEC) of the soil, the greater will be the binding of metallic cations (Nightingale 1987a). A soil's CEC is pH-dependent; therefore, the ion exchange ability of a soil to remove metals from solution is pH-dependent.

Organic complexation of a metal may enhance the metal's ability to move freely to the groundwater. Organic complexes often are stable and uncharged or negatively charged. Because of their negative or neutral charge, they are not attracted to negatively charged adsorption/ion exchange sites and are not easily removed from solution (Wilde 1994).

Precipitation of Metals

Selenium in agricultural drainage water is generally in its fully oxidized state, as selenate. Although selenites can be readily precipitated from water, even in the presence of salts, selenate precipitation is inhibited. The precipitation process relies upon the bacteria that occur naturally in the drainage water and convert the inorganic selenate to an organic complex. Part of the selenate is assimilated into the complex, part is reduced to a lower oxidation state, and the rest is reduced to zero valent selenium (Squires, et al. 1987). Further testing confirms that selenite can be readily precipitated from drainage water, even in the presence of salts, but selenate precipitation is inhibited in the presence of nitrate and sulfate (Squires, et al. 1989).

In central Florida, the dissolved iron concentrations in recharge water are much greater than in the groundwater, indicating that dissolved iron ions are being removed from the recharge water during percolation. The reduction of dissolved iron concentrations resulted from precipitation of iron or complexation into other nonsoluble species (Schiffer 1989). Iron and manganese transformations in groundwater are controlled by both the oxidation-reduction conditions and the acid-alkali balance in the water. The migration and diffusion speed of iron (Fe^{+2}) and oxygen is rather slow in water, but is accelerated by iron bacteria. The amount of iron oxidized by iron bacteria is ten times—or even hundreds of times—greater than that oxidized by the chemical reaction alone (Bao-rui 1988). In central Florida, zinc, which is more soluble than iron, was commonly found in higher concentrations in groundwater than iron (Schiffer 1989). Iron and manganese oxidation may lower the pH of a water because the oxidation reactions add acidity (as the H^+ ion) to the water. A lower pH may cause an increase in the dissolved metal concentrations in the water. Concurrent decreases in pH and Eh in the unsaturated zone may have increased the mobility of manganese beneath one stormwater impoundment in Maryland (Wilde 1994).

Dissolved Metal Concentration Increases in Groundwater

The dissolved iron concentration in the Magothy aquifer was greater after recharge than in the native groundwater and in the recharge water itself. The observed increase in iron concentration from less than 0.5 mg/L to 3 mg/L indicated a constituent of the recharge water reacting with the pyrite and marcasite (FeS_2) in the aquifer (Ragone 1977). For the first 20 ft from the well in the aquifer, the dissolved oxygen in the recharge water reacted with the pyrite to produce ferrous iron, sulfate, and hydrogen ions, according to the following equation:

$$FeS_2 + 7/2\ O_2 + H_2O \rightarrow Fe^{+2} + 2\ SO_4^{-2} + 2\ H^+$$

At distances greater than twenty feet, the dissolved oxygen concentration was zero. The reason for the increase in the dissolved iron concentration was unknown (Ragone 1977).

Although the concentrations of many trace metals were reduced through sorption/ion exchange to bottom materials and/or native sediments, elevated concentrations (greater than background) of these metals were found in groundwater beneath and downgradient of three infiltration devices in Maryland. The greater-than-expected metal concentrations could have resulted from fluctuations in pH and/or dissolved oxygen; unfavorable conditions for sorption, oxidation, ion exchange, or precipitation in the pond and/or unsaturated zone; or significantly greater flow rates through the unsaturated zone (Wilde 1994).

Mobility

In dry recharge wells in Arizona, manganese was the only metal that was mobile in the nearly neutral vadose zone sediments and the only metal to show up in the groundwater at elevated concentrations (Wilson, et al. 1990).

Aluminum mobility is governed by pH, amount of stormwater infiltration, horizontal and vertical groundwater flow, depth to water table, and existence of channels for preferential flow. Aluminum is soluble above pH 9.0. Cadmium solubility/mobility is governed by pH, redox potential, biological uptake, and solubilities of carbonates and sulfides. If sulfide is present, cadmium is nearly immobile. Cadmium complexes readily with organic and inorganic ligands but preferentially in the following order:

humic acids > carbonate > hydroxides > chlorides > sulfates

Hydrophilic and negatively charged, or neutrally charged, complexes are not likely to be retarded in the vadose zone. Copper is soluble at high pH values (greater pH than zinc or nickel) and its solubility is affected by complexation with iron and other ligands and by coprecipitation by oxides. If chromium entering an infiltration device is negatively charged, or is bound in a stable, negatively charged compound, it may move easily through the vadose zone (Wilde 1994).

Chromium was detected in groundwater in Maryland beneath stormwater infiltration devices and was not removed from solution by sorption to the sediments in the device, or to the vadose zone sediments. Lead forms compounds with hydroxides, carbonates, sulfides, and sulfates, all of which have a low solubility. Lead also is removed from solution by binding with organic matter, coprecipitating with manganese oxides, and sorbing to organic and inorganic substrates. Zinc mobility is limited by high pH, the partial pressure of carbon dioxide in the solution, and the presence of sulfide. Aqueous zinc can be reduced by coprecipitation with other minerals, cation exchange, biochemical activities, and complexation and sorption to organic and inorganic substrates (Wilde 1994).

In Maryland, groundwater pH below and downgradient of stormwater infiltration devices was less than 5.0 and tended to keep the metals in solution or solubilize metals attached to particulates. Concentrations of cadmium, chromium, and lead exceeded the U.S. Environmental Protection Agency's Maximum Contaminant Levels in some groundwater samples, and concentrations of barium, copper, nickel, strontium, and zinc in downgradient groundwater were often greater than their concentrations in native groundwater (Wilde 1994).

Experiments in Orlando, Florida (Harper 1988), concerning metal mobility in soil and its resultant stability have led to the ranking of metals in order of attenuation from recharge water:

zinc (most mobile) > lead > cadmium > manganese > copper > iron > chromium > nickel > aluminum (least mobile)

Other studies of metal pollutant mobility in soil have led to the generation of mobility classes, as shown in Tables 23 and 32 (Armstrong and Llena 1992). The mobility of these metals is seen to be much less than for most organics.

General

Table 33 summarizes the principal removal mechanisms in the soil for each metal (Crites 1985). The surface water heavy metal concentrations were the most significant variables in predicting the concentrations of the heavy metals in the groundwater (Harper 1988).

Health and Ecological Effects

Metals in groundwater can cause environmental and ecological problems, as well as human health problems. Groundwater contamination will mostly affect human consumers of the groundwater, although the discharge of contaminated groundwaters into surface waters can also have deleterious effects on aquatic organisms.

Table 32. Metal Mobility

Inorganic Pollutant	Concentration (mg/L)	Mobility Class	
		Sandy Loam	Silt Loam
arsenic	1.0	III	III
	0.01	IV	IV
cadmium	1.0	III	III
	0.01	III	IV
chromium	1.0	III	II
	0.01	IV	III
copper	1.0	IV	IV
	0.01	IV	IV
lead	1.0	IV	IV
	0.01	IV	IV
nickel	1.0	III	III
	0.01	III	III
zinc	1.0	III	III
	0.01	III	IV

Source: Armstrong and Llena 1992.

Environmental Problems

Boron concentrations in sago pondweed from several lakes and reservoirs affected by irrigation water have been found to be large enough to damage sole consumers of that pondweed, and boron is toxic, especially to citrus crops (Bouwer and Idelovitch 1987). In Wyoming, boron and selenium concentrations from local agriculture had concentrated in the wildlife in sufficient quantities to damage bird livers and eggs, to reduce fish reproduction and to kill water fowl (Peterson 1988; Deason 1989). Molybdenum is toxic to animals that forage on plants with high molybdenum concentrations (Bouwer and Idelovitch 1987). Many different heavy metals also affect aquatic organisms, including microinvertebrates, fish, and plants.

Human Health Problems

Human health problems due to metal poisoning have been linked to water supplies. Iron and manganese removal in most water supplies is primarily for aesthetic reasons (including taste). However, excess concentrations of certain elements, including arsenic, copper, lead, and selenium, in the potable water have caused disease outbreaks (Craun 1979). The following list of human health effects is summarized from Craun (1979) and Crites (1985):

- Arsenic is readily adsorbed from the gastrointestinal system and/or the lungs and distributed throughout the body, where it can bioaccumulate. Symptoms of mild chronic arsenic poisoning include fatigue and loss of energy, while symptoms of severe poisoning include gastrointestinal mucous membrane inflammation, kidney degeneration, fluid accumulation in the body, polyneuritis and bone marrow injury.
- Barium enters the body through either inhalation or ingestion. Barium ingestion can cause toxic effects on the heart, blood vessels, and nerves.

Table 33. Metal Removal Mechanisms in Soil

Element	Principal Forms in Soil Solution	Principal Removal Mechanisms
arsenic	AsO_4^{-3}	strong associations with clay fractions of soil
barium	Ba^{+2}	precipitation and sorption onto metal oxides and hydroxides
cadmium	Cd^{+2} complexes chelates	ion exchange, sorption, and precipitation
chromium	Cr^{+3} Cr^{+6} $Cr_2O_9^{-2}$ CrO_4^{-2}	sorption, precipitation, and ion exchange
cobalt	Co^{+2} Co^{+3}	surface sorption, surface complex ion formation, lattice penetration, ion exchange, chelation, and precipitation
copper	Cu^{+2} $Cu(OH)^+$ anionic forms chelates	surface sorption, surface complex ion formation, ion exchange, and chelation
iron	Fe^{+2} Fe^{+3} polymeric forms	surface sorption and surface complex ion
lead	Pb^{+2}	surface sorption, ion exchange, chelation, and precipitation
manganese	Mn^{+2}	surface sorption, surface complex ion formation, ion exchange, chelation, and precipitation
mercury	Hg^+ HgS $HgCl_3^-$ $HgCl_4^{-2}$ CH_3Hg^+ Hg^{+2}	volatilization, sorption, and chemical and microbial degradation
nickel	Ni^{+2}	surface sorption, ion exchange, and chelation
selenium	SeO_3^{-2} SeO_4^{-2}	ferric-oxide selenite complexation
silver	Ag^+	precipitation
zinc	Zn^{+2} complexes chelates	surface sorption, surface complex ion formation, lattice penetration, ion exchange, chelation, and precipitation

Source: Crites 1985.

- Cadmium, because of its similarities to zinc, will bind to sites on enzymes intended for zinc. Symptoms of mild chronic poisoning include proteinurea. Continued exposure to cadmium will lead to renal degradation, respiratory disorders such as emphysema, gastric and intestinal dysfunctions, anemia, and hypertensive heart disease. Exposures to very large quantities of cadmium have caused itai-itai disease, erythrocyte destruction, and testicular damage, but the necessarily high concentrations are unlikely to occur in groundwater. All disease outbreaks from very large doses of cadmium have resulted from direct industrial exposure.

- Chromium is necessary for glucose tolerance in animals, including man. However, large quantities of hexavalent chromium can cause tumors if inhaled, or through skin contact.
- Copper is also essential for animals, and a lack of copper can cause nutritional anemia in infants. Large doses of copper can produce vomiting and liver damage. Disease outbreaks have been traced to copper leaching from plumbing.
- Iron, common in most metal pipes, is not a desired constituent in drinking water due to taste, fixture staining, and deposit accumulation.
- Lead enters the body by either inhalation or ingestion, and accumulates in the liver, kidney, and bones. Chronic high lead exposures can cause burning in the mouth, severe thirst, vomiting, and diarrhea. Acute toxicity starts with convulsions and anemia, and may proceed to peripheral nerve disease, joint swelling, kidney degeneration, mental confusion, brain damage, and eventually death.
- Manganese, like iron, is objectionable in drinking water because it affects taste, stains fixtures, spots laundered clothes and collects in distribution systems.
- Mercury damages while it bioaccumulates in the liver, kidney, and brain. Chronic exposure results in mouth and gum inflammation, salivary gland swelling, teeth loosening, kidney damage, muscle spasms and personality changes. Acute mercury poisoning causes severe diarrhea, vomiting, kidney damage, and death. Mercury also deforms fetuses.
- Chronic selenium poisoning results in red staining of the fingers, teeth, and hair; depression; nose and throat irritation; upset stomach; and skin rashes. Acute poisoning is characterized by nervousness, vomiting, convulsions, hypertension, and respiratory failure.
- Silver poisoning results in skin, eye, and mucous membrane discoloration, and silver is retained in the body tissue indefinitely.
- Zinc, although it is necessary for life, affects the taste of water if present in excessive amounts.

GROUNDWATER CONTAMINATION ASSOCIATED WITH SALTS AND OTHER DISSOLVED MINERALS

Definition

The dissolved minerals of concern in groundwater contamination are primarily salts. These salts include compounds containing combinations of calcium, potassium, sodium, chloride, fluoride, sulfate, or bicarbonate.

Examples of Salts Contaminating Groundwater

Increasing chloride concentrations in groundwater have been used as an indicator of early groundwater contamination in Great Britain (Lloyd, et al. 1988). When using rapid infiltration for recharge, inorganic dissolved solids are of concern, and include chloride, sulfate, and sodium (Crites 1985). Sources of the dissolved solids include naturally occurring salts, landfill leachate, leaky sewerage, cesspool leachate, and other urban and agricultural sources.

Natural Occurrence

On Long Island, New York, one of the sources of dissolved solids in the recharge water was salt spray from the sea (Seaburn and Aronson 1974). Saltwater from an adjacent surface water body can infiltrate into the aquifer if the hydraulic gradient flows in that direction. A brackish lagoon near Narbonne, France has contaminated groundwater (Razack, et al. 1988).

Urban Areas

Salt application for winter traffic safety is a common practice in many northern areas, and the sodium and chloride, which are collected in the snowmelt, travel down through the vadose zone to the groundwater with little attenuation. Fertilizer and pesticide salts also accumulate in urban areas and leach through the soil to the groundwater (Merkel, et al. 1988). In Arizona, stormwater infiltration in dry wells dissolves native salts in the vadose zone that are then carried to the groundwater (Wilson, et al. 1990).

Investigations of groundwater near a landfill in Lee County, Florida, showed that the concentrations of sulfates, potassium, chloride, and sodium were 10 to 100 times greater than in the unaffected aquifer (Boggess 1975). Elevated sulfate concentrations in the groundwater beneath the city center in Narbonne, France, originated from leaking sanitary sewerage. The use of saltpeter (potassium nitrate) in the numerous wine cellars in the area also contaminated the groundwater (Razack, et al. 1988). Elevated chloride, sodium, and sulfate groundwater concentrations resulted from cesspool leachate on Long Island, New York (Smith and Myott 1975), and from septic tank leachate in Florida (Waller, et al. 1987).

Agricultural Operations

Major sources of dissolved solids in groundwaters in agricultural areas include fertilizers and pesticides. Elevated groundwater total dissolved solids concentrations of 200 mg/L and chloride and sulfate concentrations of about 10 mg/L and 20 mg/L, respectively, have been reported on Long Island, New York (Seaburn and Aronson 1974). The concentrations of sodium, chloride, and sulfate vary with the season and are likely related to precipitation and irrigation patterns (Hampson 1986).

In areas of sustained irrigation, some leaching must occur periodically to remove the salts that accumulate in the root zone of the crops (Power and Schepers 1989). Evapotranspiration concentrates the salts in the root zone. During irrigation the salts are flushed into the vadose zone and eventually into the groundwater (Schmidt and Sherman 1987). Spray irrigation with secondary treated effluent can increase chloride concentrations and specific conductance in shallow aquifers (Brown 1982). Grazing cattle return 75 to 80% of the of the potassium in their forage to the soil (Reichenbaugh 1977).

High groundwater salinity has been noted in the San Joaquin Valley, under the agricultural areas near Fresno, California, in irrigated areas of Arizona and New Mexico, where the salinity has been increasing since the 1930s, and in the Corn Belt and Great Lakes States (Schmidt and Sherman 1987; Deason 1989; Bouwer 1989; Sabol, et al. 1987; and Mossbarger and Yost 1989). Most salts in groundwater below irrigated areas have resulted from the leaching of natural salts from the arid soils. Use of sludge as fertilizer on sandy loam soil in New Jersey increased the total dissolved solids concentration of the groundwater (Higgins 1984). On Long Island, New York, recharge of the groundwater has led to an increase in the sodium and chloride concentrations above the background concentrations (Schneider, et al. 1987). Mathematical modeling has led to the conclusion that salt from agricultural return flows is the greatest single contributor (about 40%) of salt to Upper Midwest groundwater (Schmidt and Sherman 1987).

Salt Removal Processes in Soils

Most salts are not attenuated during movement through soil. In fact, salt concentrations typically increase due to leaching of salts out of soils. Groundwater salt concentration decreases may occur with dilution of less saline recharging waters. Use of lower-salinity water as recharge water at the Leaky Acres stormwater recharge facility in Fresno, California, was shown to decrease the salt concentrations in the groundwater (Nightingale and Bianchi 1977a). Reduction in the pH of groundwater, such as would result from nitrification and the biodegradation of carbonaceous substances, resulted in the dissolution of soil minerals and subsequent increases in the total dissolved solids concentrations and the hardness of groundwater at the Whittier Narrows site in Los Angeles County, California (Nellor, et al. 1985). This effect was noted in Florida during the deep-well injection of acidic, high–oxygen demanding industrial waste. At first, neutralization of the waste occurred through solution of the calcium carbonate in the limestone. Later, the calcium concentration in the aquifer increased and the pH decreased, but the effects have still been confined to the lower strata of the Floridan aquifer (Goolsby 1972).

Leaching of Salts

Salt leaching is a greater concern in arid areas of the United States because the irrigation requirements for the arid areas are great and the irrigation water collects the salts that have been concentrated in the soil by increased evapotranspiration. Salts that are still in the percolation water after it travels through the vadose zone will contaminate the groundwater (Sabol, et al. 1987; Bouwer 1987). The rate of contaminated water movement in a water-table aquifer is highest when the water table is highest (Boggess 1975).

Nonuniform irrigation and preferential flow allow the percolation water to reach the groundwater much faster and reduce the amount of salt removal that could occur during slower water passage through the soil (Bouwer 1987), especially for the chloride ion, which is not readily adsorbed. Irrigation efficiency and interval significantly affect groundwater chloride concentrations. For example, fields in Arizona and New Mexico that were irrigated at nearly 100% efficiency had much higher chloride concentrations in the root zone of the soil (Sabol, et al. 1987). The higher the salt concentration of the soil solution, the higher the soil hydraulic conductivity will be for a given sodium adsorption ratio (SAR) (Bouwer and Idelovitch 1987). Schmidt and Sherman (1987) found a direct relationship between concentrations of groundwater nitrates and salts.

Solubility Equilibrium of Salts

An equilibrium exists between the recharge water and the high groundwater for calcium and total dissolved solids. For chloride, sodium, and sulfate, reductions in concentrations entering the recharge system are likely accounted for by differences in seasonal precipitation, with a higher loading in the summer than in the winter. Changes in the groundwater concentration reflect these loading differences (Hampson 1986). Potassium exchanges with hydrogen ions on the clay during percolation. Other exchanges cause the calcium and magnesium concentrations to be much greater than had been predicted (Ragone 1977). Deep-well injection waters have shown an increase in alkalinity and bicarbonate concentrations, reflecting the mineralization of the organic compounds. Dissolved calcium and bicarbonate are the primary products of limestone dissolution. Many parameters in natural groundwater systems are controlled, or are influenced, by the calcium carbonate equilibrium system.

Removal of Salts in Soil

Soil is not very effective at removing most salts. Depth of dissolved mineral penetration in soil has been studied (Close 1987) at a site with a shallow, unconfined aquifer. This study found that sulfate and potassium concentrations decreased with depth, while sodium, calcium, bicarbonate, and chloride concentrations increased with depth in the soil. The dissolution of the aquifer material may be the source of many of the chloride, bicarbonate, calcium, and sodium ions. The same increase in salt concentrations with depth was noted in the agricultural (irrigated) areas of Arizona and New Mexico, and may be a result of little water and salt uptake by the plants. On Long Island, New York, it was noted that the heavy metals load was significantly reduced during passage through the soil, while chloride was not reduced significantly. This indicated that the soil does not contain an effective removal process for chloride salts (Ku and Simmons 1986). However, fluorine is removed in soil through sorption and precipitation processes (Crites 1985).

In another study (Waller, et al. 1987), it was found that chloride and sulfate concentrations from septic tanks increased with depth below the leach field, but were rapidly diluted downgradient. The primary controls on the leachate movement are the lithology and layering of the geologic materials, hydraulic gradient slopes, and the volume and type of use the septic system receives. Dilution occurs more rapidly in limestone than in sand.

Once contamination with salts begins, the movement of salts into the groundwater can be rapid. The salt concentration may not lessen until the source of the salts is removed. The cations sodium, potassium, calcium, and magnesium appeared in a shallow aquifer three to six months after the source water was applied to the soil (Higgins 1984).

At three stormwater infiltration locations in Maryland, the nearby use of deicing salts and their subsequent infiltration to the groundwater shifted the major-ion chemistry of the groundwater to a chloride-dominated solution. Although road deicing occurred only three to eight times a year, increasing chloride concentrations were noted in the groundwater throughout the three-year study, indicating that groundwater systems are not easily purged of conservative contaminants, even if the groundwater flow rate is relatively high. Sodium and/or calcium concentrations also were constantly elevated in the groundwater beneath and downgradient of the infiltration devices (Wilde 1994).

Control of groundwater salt contamination in agricultural areas should result from maintaining irrigation rates at near the minimum leaching rate (Bouwer 1987). This control, with other crop management practices, will slow the leaching rate of the salt to the groundwater (Lee 1990). Reduction of the dissolved solids concentrations will decrease by dilution through recharge with "cleaner" water (Nightingale, et al. 1983).

Health and Other Effects

Besides direct effects, dissolved solids concentrations can also affect the pH of the water. Problems may then be caused by the effect of pH on high concentrations of undesirable elements, such as iron, manganese, and aluminum in acid waters, and sodium, carbonates, and bicarbonates in alkaline waters (Bouwer 1987).

Human Health Problems

High total dissolved solids concentrations are objectionable because of possible physiological effects, mineral taste, and corrosion. High concentrations of chloride ions in water affects its taste and accelerates the corrosion of pipes and household appliances. Excess sodium can cause health problems, especially for people who are on sodium-restricted diets, due either to hypertension, edema from congestive cardiac failure, and toxemia in pregnant women. High concentrations of sulfate ions affect the taste of the water and are a laxative to humans (Crites 1985). Although small quantities of fluoride are recommended in water by the American Dental Association, excessive amounts of fluoride can mottle teeth, instead of protecting them.

Agricultural Problems

Sodium can adversely affect crops by causing leaf burn in almonds, avocados, and stone fruits. Bicarbonate in spray-applied irrigation will leave an unappealing white residue after the water evaporates. Sulfate affects the growth of many plants, and can cause leaf burn in still others (Bouwer and Idelovitch 1987).

GROUNDWATER CONTAMINATION ASSOCIATED WITH SUSPENDED SOLIDS

Definition

The term "suspended solids" is usually defined as the nonfilterable portion of a water sample after filtering through a 0.45-μm filter. This definition does not accurately describe the fate of this material. Depending on specific gravity and particle sizes, the "suspended" material also includes floatable matter, rapidly settleable matter, and matter that could settle out of suspension over highly variable periods of time.

Soil Removal Processes of Suspended Solids

Suspended solids are of concern because of the potential for clogging the infiltration area (Crites 1985). The recharge water should be of low salinity and turbidity (Nightingale and Bianchi 1977b). When the groundwater salinity was reduced by using less-salty recharge water in Fresno, California, the turbidity of the groundwater increased. Leaching of poorly crystallized and extremely fine colloids from the soil and into the groundwater had occurred. This effect was only temporary—with the groundwater returning to its original turbidity soon after recharge began—and was not observed outside of the recharge basin area. Changes in the recharge water salinity, however, may cause this effect to return (Nightingale and Bianchi 1977b).

Laboratory studies on the movement of fine particulate matter through sand aquifers found that the movement is controlled first by the nature of the particle, second by the cation and anion concentrations of the percolating water, and third by the pore size distribution of the soil to the aquifer (Nightingale and Bianchi 1977b). As water flows through passages formed by the soil particles, suspended and colloidal particles too small to be retained at the surface are thrown off their streamline through hydrodynamic actions, diffusion, impingement, and sedimentation. The particles may then be adsorbed onto stationary soil particles. The degree of trapping and adsorption of suspended solids by soils is a function of the suspended solids concentration and size distribution, soil characteristics, and hydraulic loading (EPA 1992). The soil profile will filter out suspended solids from recharge water, and once the particles are in the soils, biological and chemical degradation may occur. Fine- to medium-textured soils remove essentially all suspended solids from the wastewater by straining (Bouwer 1985), while coarse-textured soils enable deeper penetration of suspended and colloidal particles in the soil (Treweek 1985).

DISSOLVED OXYGEN GROUNDWATER PROBLEMS

Dissolved oxygen (DO) in any water is controlled not only by how much DO exists and how much can be produced by aquatic plants, but also by the oxygen needs of the other organisms and compounds in the water. As groundwater recharging proceeds, dissolved oxygen is depleted due to the proliferation of iron bacteria and *T. thioparus*, which leads to anaerobic conditions in the groundwater (Bao-rui 1988). Increases in both dissolved oxygen and temperature should increase virus inactivation (Jansons, et al. 1987b). Groundwater has no natural reaeration process available, so once depleted, groundwater DO will remain very low.

COD levels from injection well wastes were reduced approximately 80% within relatively short distances from an injection point near Pensacola, Florida (Ehrlich, et al. 1979b). During recharge of the Magothy aquifer at Bay Park, New York, it was found that dissolved oxygen persisted for about 12 ft away from the recharge well. At greater distances, the water is essentially oxygen free. Two occurrences can account for the oxygen loss from the recharge water as it moves through the aquifer. First, oxygen reacts with pyrite in the formation to produce ferrous iron, sulfate, and hydrogen. Second, microbial respiration associated with waste stabilization depleted the oxygen supply (Ehrlich, et al. 1979a).

Chapter 4

TREATMENT BEFORE DISCHARGE OF STORMWATER

One of the best overall urban runoff control strategies may be to encourage infiltration of stormwater to replace the natural infiltration capacity lost through urbanization. This significantly reduces the volume of runoff discharged to surface waters, including pollutants. This strategy also improves groundwater conditions by reducing the lowering rate of urban water tables. Exfiltration from groundwater into local streams during dry periods can also substantially improve receiving water biological conditions. The EPA (1983) concluded, as part of the Nationwide Urban Runoff Program, that stormwater can be safely infiltrated to groundwater, if done carefully. Issues that must be considered include a knowledge of pollutant concentrations from different areas, pollutant removals in the vadose zone, and necessary pretreatment that may be needed before infiltration. This chapter reviews characteristics of urban runoff pollutants that will affect their fates in treatment processes, along with reported performance of stormwater treatment devices.

SOLUBILITIES AND TREATMENT POTENTIALS OF SIGNIFICANT URBAN RUNOFF TOXICANTS

This chapter discusses chemical reactions, solubilities and fates of significant urban runoff pollutants. The information presented here is based upon a review of the urban runoff and environmental chemistry literature, and addresses toxic heavy metals and organic pollutants that have been detected in various urban runoff waters. This information can be used to identify the potential removal mechanisms that may be available in stormwater control practices, and to identify the potential transport and fate mechanisms of the pollutants in the surface or subsurface receiving waters.

Arsenic

Arsenic Sorption

Arsenic can be adsorbed onto clays, iron oxides, and inorganics (Callahan, et al. 1979).

Arsenic Fate/Treatment

Callahan, et al. (1979) stated that many environmental fate mechanisms, except for photolysis, can be important for arsenic. Arsenic can either remain suspended or accumulate in sediments (Callahan, et al. 1979). Phillips and Russo (1978) stated that arsenic may be bacterially methylated, much like mercury, to form highly toxic methyl-arsenic or dimethylarsenic. These methylated forms of arsenic are very volatile and are readily oxidized to less toxic forms.

Cadmium

Cadmium Filterable Fraction, Solubility, and Sorption

About 40 to 50% of the cadmium in roof, loading docks, and street runoff sampled by Pitt, et al. (1995) was found in filtered sample components, while the other storm-water source areas all had less than 20% associated with the filtered sample component. Pitt and Amy (1973) studied the leachability of cadmium from street dirt, along with other metals, and found that in typical urban runoff concentrations, leachable cadmium values of less than 1 µg/L occurred in moderately hard water after an exposure of 25 days. This leachable fraction was 14% of the total cadmium in the mixture. Wilber and Hunter (1980), in an urban receiving water study in Lodi, New Jersey, found that with most low flows in the Saddle River, the cadmium was mostly dissolved. However, during wet-weather conditions, most of the cadmium was associated with undissolved particulates. Callahan, et al. (1979) stated that adsorption of cadmium onto organics, clays, hydrous iron, and manganese oxides is important in polluted water.

Chromium

Chromium Filterable Fraction

Filtered stormwater samples generally contained less than 10% of the total chromium detected by Pitt, et al. (1995). Pitt and Amy (1973) found that the leachable fraction of chromium associated with street dirt in moderately hard water was about 4 µg/L, or about 0.3% of the total chromium in the mixture.

Copper

Copper Filterable Fraction, Sorption, and Solubility

The filtered stormwater samples analyzed by Pitt, et al. (1995) generally had less than 20% of the total copper concentrations. Wilber and Hunter (1980), in a study of an urban river in Lodi, New Jersey, found that the readily available copper (at a pH of about 7) was about 13% of the street dirt and runoff solids total copper content. Pitt and Amy (1973) found that the leachable fraction of copper associated with street dirt was about 160 µg/L, or about 36% of the total copper in the mixture, with moderately hard water conditions. The adsorption of copper can reduce its mobility and enrich suspended and settled sediments (Callahan, et al. 1979). Copper is absorbed onto organics, clay minerals, hydrous iron, and manganese oxides.

Iron

Iron Filterable Fraction, Sorption, and Solubility

Pitt and Amy (1973) found that the leachable fraction of iron in street dirt was about 50 µg/L, or much less than 1% of the total iron in a mixture with moderately hard water. They also stated that the principal inorganic iron forms, near pH 7, are iron oxide, hydroxide, sulfate, nitrate, and carbonate. Phillips and Russo (1978) stated that the soluble ferrous form of iron (Fe^{+2}) is readily oxidized to the insoluble ferric, or trivalent (Fe^{+3}) state in most natural surface waters. A substantial fraction of iron in natural waters is therefore associated with suspended solids.

Lead

Lead Filterable Fraction, Sorption, and Solubility

The filtered stormwater samples analyzed by Pitt, et al. (1995) generally had less than 20% of the total lead concentrations. The EPA (1976) stated that most lead salts are of low solubility. The aqueous solubility of lead ranges from 500 µg/L in soft water to 3 µg/L in hard water (EPA 1976). Durum (1974) stated that lead carbonate and lead hydroxide are soluble lead forms at pH values of 6.5 or less, with low alkalinity conditions (less than 30 mg/L alkalinity as $CaCO_3$). The soluble lead concentrations under these conditions can reach 40 to several hundred µg/L. If the alkalinity is greater than 60 mg/L, and if the pH is near 8, however, the dissolved lead will be less than 10 µg/L. Callahan, et al. (1979) stated that lead carbonate and lead sulfate control lead solubility under aerobic conditions and normal pH values. Lead sulfide and lead ions, however, control lead solubility in anaerobic conditions. In polluted water, the organic complexes of lead are most important in controlling lead solubility. Phillips and Russo (1978) stated that most lead is probably precipitated in natural waters due to the presence of carbonates and hydroxides.

Pitt and Amy (1973) found that the leachable fraction of lead in a street dirt and water mixture was about 40 µg/L, or about 3% of the total lead, in moderately hard water. Wilber and Hunter (1980) found that the readily available fraction of lead was about 20% of the total lead in street dirt and runoff solids. They also found that under most low flow river conditions, most of the lead was dissolved, but under wet-weather conditions, most of the lead was insoluble. Solomon and Natusch (1977) also examined the solubilities of lead associated with street dust. They found solubilities ranging from 500 to 5000 µg/L, which was 0.03 to 0.3% of the initial mixture total lead concentration. However, the test mixture of street dirt with water was very high (1750 mg/L lead). Rolfe and Reinhold (1977) found that about 80% of the lead in stream water was insoluble and associated with suspended solids.

Nickel

Nickel Filterable Fraction

Very few of the filtered stormwater samples analyzed by Pitt, et al. (1995) had detectable (>1 µg/L) nickel concentrations. Wilber and Hunter (1980) found that the readily available nickel fraction of street dirt and runoff solids was about 4% at close to neutral pH conditions. Pitt and Amy (1973) found that the leachable fraction of nickel associated with street dirt, in a moderately hard water mixture, was about 30 µg/L, or about 7% of the total nickel in the mixture.

Mercury

Mercury Fate/Treatment

Callahan, et al. (1979) stated that almost all of the environmental processes are important when determining the fate of mercury in aquatic environments. Phillips and Russo (1978) reported that inorganic mercury concentrations, availability of inorganic mercury, pH, microbial activity, and redox potential all affect mercury methylation rates. In general, more methylmercury is produced when more inorganic mercury is present. Chemical agents which precipitate mercury, such as sulfide, reduce the availability of mercury for methylation, but only when present in large quantities. At neutral pH values, the primary product of mercury methylation is monomethylmercury. Methylation can occur under both aerobic and anaerobic conditions, but more mercury is produced when more bacteria are present.

Zinc

Zinc Filtered Fraction, Solubility, and Sorption

In contrast to most other heavy metals, filtered stormwater samples contain most of the total zinc concentrations observed, except for storage area and vehicle service area runoff (Pitt, et al. 1995). The major zinc source in urban areas is probably galvanized metal. Short contact periods of naturally acidic rainwater with roof galvanized metal flashings and gutters (or other galvanized metal in the drainage area) result in elevated dissolved zinc concentrations. The flow time for these waters to the outfall is rather short, and the zinc generally remains dissolved. In the receiving waters, more time is available to form precipitates or associations with particulates. Other sources of zinc are related to automobile tire wear and exhaust emissions, which are in particulate forms, and are generally insoluble.

Durum (1974) stated that the solubility of zinc is less than 100 μg/L at pH values greater than 8, and less than 1,000 μg/L for pH values greater than 7, if there is a high concentration of dissolved carbon dioxide. Phillips and Russo (1978) stated that zinc sulfates and halides are soluble in water, but zinc carbonates, oxides, and sulfides are insoluble. Wilber and Hunter (1980), in a study of an urban stream near Lodi, New Jersey, found that the readily available zinc in street dirt and runoff solids was about 17% of the total zinc. Most of the zinc in the river during low flow conditions was dissolved, while during wet weather it was mostly in the solid form. Pitt and Amy (1973) found that the leachable fraction of zinc was about 170 μg/L, or about 8% of the total street dirt zinc, in a moderately hard water mixture.

Phenols and Chlorophenols

Phenols and Chlorophenol Filtered Fraction

Callahan, et al. (1979) stated that the solubility of chlorinated phenols in water solutions is low, but increases when the pH increases. Phenoxide salts are also more soluble than the corresponding phenol in water with neutral pH conditions.

Phenols and Chlorophenol Fate/Treatment

Phenol may be biochemically hydroxylated to ortho- and paradihydroxybenzenes and readily oxidized to the corresponding benzoquinones (EPA 1979). These may in turn react with numerous components of industrial waters, sewerage, or other waste streams such as mercaptans, amines, or the -SH, or -NH group of proteins. Phenol has also been shown to be highly reactive to chlorine in dilute solutions over a wide pH range. The chlorination of phenol to toxic chlorophenols has been demonstrated under conditions similar to those used for disinfection of wastewater effluent.

Pentachlorophenol (PCP)

Pentachlorophenol Filtered Fraction

PCP is slightly soluble in water, while PCP salts are highly soluble in water (Callahan, et al. 1979).

Pentachlorophenol Fate/Treatment

PCP can undergo photochemical degradation in solutions in the presence of sunlight, with subsequent formation of several chlorinated benzoquinones (EPA 1979).

Sodium-PCP can be decomposed directly by sunlight with the formation of numerous products. Microorganisms have also been reported to metabolize PCP. PCP has also been reported to persist in warm and moist soils for a period of one year.

2,4-Dimethylphenol (2,4-DMP)

2,4-Dimethylphenol Filtered Fraction

2,4-DMP is slightly soluble in water (Callahan, et al. 1979).

General Polycyclic Aromatic Hydrocarbons (PAHs)

PAH Filtered Fraction

PAHs are basically insoluble in water (Callahan, et al. 1979).

PAH Fate/Treatment

These materials will be adsorbed onto suspended particulates and biota. The dissolved portion of these compounds can undergo direct photolysis at a rapid rate. Biodegradation and biotransformation by benthic organisms of PAH-contaminated sediments is believed to be their ultimate fate (Callahan, et al. 1979). Because of the low solubility of PAHs in water, biological treatment has little benefit. However, because of the attraction of PAHs to solids, physical solids separation processes can be very effective in reducing PAH concentrations (PHS 1981).

Benzo (a) Anthracene

Benzo (a) Anthracene Filtered Fraction

No detectable (>1 μg/L) benzo (a) anthracene was found by Pitt, et al. (1995) in filtered stormwater sample fractions. The solubility of benzo (a) anthracene in water is about 10 to 45 μg/L (Verschueren 1983).

Benzo (a) Anthracene Fate/Treatment

More than half of the benzo (a) anthracene was adsorbed onto waterborne particulates (including aggregates of dead plankton and bacteria) after just 3 hr of exposure (Verschueren 1983). Physical treatment of sewage can reduce the benzo (a) anthracene concentrations by about 80%, while biological treatment can remove almost all of the benzo (a) anthracene, leaving less than 0.1 μg/L in the effluent. Ozonation reduced the benzo (a) anthracene concentrations in sewage effluent by about 95%, while chlorination reduced the concentrations by about 50%.

Benzo (b) Fluoranthene

Benzo (b) Fluoranthene Fate/Treatment

Physical sewage treatment processes reduced benzo (b) fluoranthene concentrations by 50 to 80%, while biological processes provided almost complete removal (Verschueren 1983). Chlorination alone accounted for about a 33% reduction. Water treatment reduced initial 0.15 µg/L benzo (b) fluoranthene concentrations by about 70%. Sedimentation in a storage reservoir only slightly reduced the concentrations.

Benzo (k) Fluoranthene

Benzo (k) Fluoranthene Fate/Treatment

Physical sewage treatment reduced concentrations of benzo (k) fluoranthene from 8 to about 2 µg/L (Verschueren 1983). Biological treatment further reduced the concentrations to less than 0.1 µg/L. Chlorination alone reduced the concentrations by about 60%, from an initial value of about 70 µg/L.

Benzo (a) Pyrene

Benzo (a) Pyrene Filtered Fraction

Benzo (a) pyrene's solubility is about 3 µg/L (Verschueren 1983).

Benzo (a) Pyrene Fate/Treatment

Benzo (a) pyrene can be degraded in soil that is inoculated with special bacteria, with as much as 80% destroyed after 8 days (Verschueren 1983). In natural estuarine waters, its degradation rate is only about 2 µg/L destroyed per 1,000 days. Its volatilization half-life is about 1,000 hr (40 days) in waters moving about 1 m/sec with winds of about 2 m/sec. The volatilization half-life extends to about 10,000 hours (400 days) for still water and calm air, and decreases to about 400 hours (20 days) for very violent mixing conditions. About 70% of a benzo (a) pyrene mixture, having an initial concentration of 3 µg/L, was adsorbed onto particles after 3 hr.

From 90 to 99% removal of benzo (a) pyrene was found using activated carbon water treatment in waters having initial concentrations of 5 to 50 mg/L (Verschueren 1983). Chlorination (6 mg/L Cl_2) also reduced initial concentrations of 50 mg/L benzo (a) pyrene by 98%. Physical wastewater treatment reduced benzo (a) pyrene concentrations by about 65 to 95%, and biological treatment further reduced these concentrations by another 50 to 99%.

Fluoranthene

Fluoranthene Filtered Fraction

The observed median filterable portion of fluoranthene in stormwater (in the range of 0.5 to 14 µg/L) was about 85% of the total sample concentration (Pitt, et al. 1995). The water solubility of fluoranthene is about 200 µg/L (Harris 1982).

Fluoranthene Fate/Treatment

Harris (1982) reported that sedimentation processes were the most important removal mechanism for fluoranthene, with removals of about 65%. Biological treatment increased the removal to about 95%. Verschueren (1983) also reported that physical sewage treatment processes reduced initial fluoranthene concentrations of 3 to 45 µg/L by about 60%, and biological treatment further reduced the fluoranthene by another 80%. Water treatment reduced the raw water fluoranthene concentrations of 0.15 mg/L by about 50% using filtration, and by another 50% by chlorination. Storage in a reservoir reduced the fluoranthene concentrations by less than 10%.

Naphthalene

Naphthalene Solubility and Filtered Fraction

The observed median filterable portion of naphthalene (in the range of 7 to 82 µg/L) in runoff samples was about 25% (Pitt, et al. 1995). At about 32 mg/L, the solubility of naphthalene is quite high compared to other PAHs (Howard 1989). Naphthalene is moderately adsorbed by soils and sediments, but to a much less extent than for other PAHs. It is weakly sorbed by sandy soils, and tests have found that less than 1% was sorbed by particulate matter in a variety of surface waters (Howard 1989).

Naphthalene Fate/Treatment

In rapidly flowing streams, volatilization accounted for about 80% and sediment adsorption accounted for about 15% of the removal of naphthalene from the water column (Howard 1989). In deeper and slower-moving water, biodegradation (having a half-life of about 1 to 9 days) was probably the most important fate mechanism. Adsorption onto sediments is probably only a significant removal mechanism in waters having high solids concentrations and slow-moving waters, such as in lakes. Photolysis degrades naphthalene in surface waters with a half-life of about 3 days, but is much less efficient in deeper waters. In 5-m-deep water, the photolysis half-life was about 550 days. The presence of algae can substantially increase the photolysis rate of naphthalene.

Howard (1989) reported that naphthalene in water biodegrades after a short acclimation period. Biodegradation of sediment-bound naphthalene is 8 to 20 times faster than in water. In heavily contaminated sediment, the biodegradation half-life is about 5 hr, but can be longer than 3 months in less contaminated sediments. No anaerobic biodegradation of naphthalene in laboratory tests was observed after 11 weeks. The evaporation half-life of naphthalene in surface waters is about 5 hr for moderate current and wind conditions. The expected half-life of naphthalene in surface waters due to evaporation losses is expected to be about 50 hr in rivers and 200 hr in lakes. Microbial degradation rates were about 0.1 µg/L per day. Less than 1% of the naphthalene was sorbed to particles in water after 3 hr of exposure. Ion exchange water treatment was close to 100% effective, and the evaporation half-life of naphthalene was reported to be about 7 hr at a water depth of 1 m. Naphthalene would be readily removed by physical and biological treatment processes.

Phenanthrene

Phenanthrene Filtered Fraction

Its solubility in water is relatively high for a PAH, being about 1,000 μg/L (Verschueren 1983). Pitt, et al. (1995) did not detect any filterable phenanthrene in stormwater above the detection limit (about 1 μg/L).

Pyrene

Pyrene Filtered Fraction

The observed median filterable portion of pyrene (about 1 to 19 μg/L) in stormwater samples was about 95% (Pitt, et al. 1995). Its solubility in water is about 160 μg/L (Verschueren 1983).

Pyrene Fate/Treatment

Pyrene can be photodegraded from soils by UV radiation (Verschueren 1983). Chlorination at 6 mg/L chlorine for 6 hours decreased initial pyrene concentrations of 27 mg/L by about 25% (Verschueren 1983). Physical wastewater treatment processes decreased pyrene concentrations by about 80%, and biological processes further decreased the pyrene concentrations by about 98%. Reservoir storage of river water decreased pyrene concentrations by about 25%. Filtration further decreased the concentrations by another 40%, and chlorination further decreased the pyrene concentrations by another 60%.

Chlordane

Chlordane Filtered Fraction

Chlordane's solubility in water is about 60 μg/L (Verschueren 1983). Pitt, et al. (1995) did not find any chlordane (detection limit of about 0.3 μg/L) in the filtered portion of stormwater samples.

Chlordane Fate/Treatment

The persistence of chlordane in water in sealed jars exposed to sunlight indicated a 15% decrease after 8 weeks. Chlordane was reduced by 75 to 100% from soils after 3 to 5 years (Verschueren 1983).

Butyl Benzyl Phthalate (BBP)

Butyl Benzyl Phthalate Filtered Fraction

The only observed filterable value of butyl benzyl phthalate (BBP) (16 µg/L) detected my Pitt, et al. (1995) in stormwater was 33% of the total value. BBP's solubility in water is about 3 mg/L (Verschueren 1983).

Butyl Benzyl Phthalate Fate/Treatment

BBP does undergo biodegradation with relatively complete removals within one month (Verschueren 1983). Biodegradation using activated sludge from a wastewater treatment plant was reported to be 99% effective after 48 hr. Biodegradation in natural river waters was about 80% effective after 1 week of exposure. Photodegradation and chemical degradation (through hydrolysis) of BBP is much less effective, with reported half-lives of greater than 100 days.

Bis (2-Chloroethyl) Ether (BCEE)

Bis (2-Chloroethyl) Ether Filtered Fraction

The two observed filterable fractions of bis (2-chloroethyl) ether (BCEE) (17 and 23 µg/L) found by Pitt, et al. (1995) in stormwater were 19 and 50% of the concentrations observed in the unfiltered samples. BCEE solubility in water is about 1 mg/L (Howard 1989). It is also adsorbed at low values onto fine sand, implying that it would be highly mobile in soils and could leach rapidly to groundwaters.

Bis (2-Chloroethyl) Ether Fate/Treatment

Bis (2-chloroethyl) ether (BCEE) may degrade in soils, but acclimation may be necessary (Howard 1989). The volatilization half-life of BCEE in streams was estimated to be about 4 days, while the volatilization half-life of BCEE in lakes was estimated to be about 180 days. Photolysis is not expected to be important, but biodegradation can reduce BCEE concentrations by 50% over 35 days. After acclimation, only 9 days were required to remove 50% of the BCEE by biodegradation. Conventional water treatment removed about 80% of the BCEE, while activated carbon, when added to conventional water treatment processes, removed all of the BCEE (Verschueren 1983).

Bis (2-Chloroisopropyl) Ether (BCIE))

Bis (2-Chloroisopropyl) Ether Filtered Fraction

The solubility of bis (2-chloroisopropyl) ether (BCIE) was reported to be 1700 mg/L (Verschueren 1983). No concentrations greater than the detection limit of about 1 µg/L were found by Pitt, et al. (1995) in filtered stormwater samples.

Bis (2-Chloroisopropyl) Ether Fate/Treatment

Basu and Bosch (1982), in their summary of the literature concerning bis (2-chloro-isopropyl) ether (BCIE), reported that hydrolysis is probably its most significant trans-formation process in aquatic systems. The overall half-life of BCIE was estimated to vary between 3 and 30 days in rivers and 30 to 300 days in lakes and groundwaters. The evaporation half-life in surface waters was estimated to be similar to the hydrolysis half-life. Leaching of BCIE is expected to be important in soils. They also reported that BCIE is unlikely to be significantly sorbed by plants.

Activated carbon treatment of contaminated water resulted in almost complete re-moval of BCIE. Conventional water treatment reduced the BCIE water content from 24 µg/L to below detection limits (Verschueren 1983).

1,3-Dichlorobenzene (1,3-DCB)

1,3-Dichlorobenzene Filtered Fraction and Sorption

The observed median filterable portion of 1,3-DCB (3 to 47 µg/L) found by Pitt, et al. (1995) in stormwater was about 75% of the unfiltered sample concentrations. The solubility of 1,3-DCB is about 125 mg/L (Verschueren 1983). 1,3-DCB may be mod-erately to tightly adsorbed to soils, but leaching can occur (Howard 1989).

1,3-Dichlorobenzene Fate/Treatment

Bacterial degradation disturbed the chemical ring structure of 1,3-dichlorobenzene (1,3-DCB) within 96 hours (Verschueren 1983). Biotransformation is likely the most significant transformation process, with a half-life of about 580 days in a river system. Sedimentation and volatilization processes decrease 1,3-DCB concentrations by half over about 1.5 days in rivers and 50 days in lakes. Biodegradation under aerobic con-ditions and volatilization from soil may be important (Howard 1989). Adsorption of 1,3-DCB to sediment is a major environmental fate mechanism. 1,3-DCB is also quite volatile from water, with a half-life of about 4 hours in moderately turbulent streams. It may biodegrade under aerobic conditions in water, but is not expected to degrade under anaero-bic conditions (such as in polluted sediments). Hydrolysis, oxidation, and direct pho-tolysis are not expected to be important fate mechanisms of 1,3-DCB in the aquatic environment.

Summary

Most of the organics and metals are associated with the nonfilterable (suspended solids) fraction of the wastewaters during wet weather. Exception were for stormwater zinc, fluoranthene, pyrene, and 1,3-dichlorobenzene, which were found mostly (>50%) in the filtered sample portions. However, dry-weather wastewater flows tended to be much more associated with dissolved sample fractions.

Many processes will affect these pollutants. Sedimentation is the most common fate and control mechanism for particulate-related pollutants. This would be common for most stormwater pollutants. Exceptions include the four stormwater constituents noted above, which were mostly associated with the filterable sample portions. Particulate removal can occur in many control processes, including catchbasins, screens, drainage systems, and detention ponds. These control processes allow removal of the accumulated polluted sediment for final disposal in an appropriate manner.

Tables 34 and 35 summarize the likely fate mechanisms for these compounds (Callahan, et al. 1979). Biological or chemical degradation of the toxicants may occur, but is quite slow for many of the pollutants in anaerobic environments. Degradation of the soluble pollutants in the water column may occur, especially when near the surface in aerated waters. Volatilization is also a mechanism that may affect many of the detected organic toxicants. Increased turbulence and oxygen supplies would encourage these processes that may significantly reduce pollutant concentrations. Sorption of pollutants onto solids and metal precipitation increase the sedimentation potential of the pollutants and also encourage more efficient bonding of the pollutants in soils, preventing their leaching to groundwaters.

OUTFALL PRETREATMENT OPTIONS BEFORE STORMWATER INFILTRATION

Sedimentation Treatment

Wet Detention Ponds

Detention ponds are probably the most common management practice for the control of stormwater runoff. If properly designed, constructed, and maintained, wet detention ponds can be very effective in controlling a wide range of pollutants and peak runoff flow rates.

There are many kinds of detention ponds, including dry ponds (which typically contain no water between storms), wet ponds (which contain standing water between storms), and combination ponds (which drain slowly after storms and may contain a small permanent pool). In a partial survey of cities in the U.S. and Canada, the American Public Works Association found more than 2,000 wet ponds (about half of which were publicly owned), more than 6,000 dry ponds, more than 3,000 parking lot multi-use detention areas, and more than 500 rooftop storage facilities (Smith 1982).

In selected areas of the U.S., detention ponds have been required for some time and are therefore much more numerous than elsewhere. In Montgomery County, Maryland, as an example, detention ponds were first required in 1971, with more than 100 facilities planned during that first year, and about 50 actually constructed. By 1978, more than 500 detention facilities had been constructed in Montgomery County alone (Williams 1982). In DuPage County, Illinois, near Chicago, more than 900 stormwater detention facilities (some natural) receive urban runoff (McComas and Sefton 1985).

The Nationwide Urban Runoff Program (NURP) included full-scale monitoring of nine wet detention ponds (EPA 1983). The Lansing, Michigan, project included two greatly enlarged pipe sections within the storm sewerage system (up-sized pipes) plus a

Table 34. Importance of Environmental Processes for the Aquatic Fates of Various Polycyclic Aromatic Hydrocarbons and Phthalate Esters

Environmental Process[a]	Anthracene	Fluoranthene	Phenanthrene	Diethyl Phthalate (DEP)	Di-n-Butyl Phthalate (DBP)	Bis (2-Ethyl-hexyl) Phthalate (DEHP)	Butyl Benzyl Phthalate (BBP)
Photolysis	dissolved portion may undergo rapid photolysis	dissolved portion may undergo rapid photolysis	dissolved portion may undergo rapid photolysis	not important	not important	not important	not important
Volatilization	may be competitive with adsorption	may be competitive with adsorption	may be competitive with adsorption	not important	not important	not important	not important
Sorption	adsorbs onto suspended solids; movement by suspended solids is important transport process	adsorbs onto suspended solids; movement by suspended solids is important transport process	sorbed onto suspended solids; movement by suspended solids is important transport process	sorbed onto suspended solids and biota; complexation with humic substances most important transport process	sorbed onto suspended solids and biota; complexation with humic substances most important transport process	sorbed onto suspended solids and biota; complexation with humic substances most important transport process	sorbed onto suspended solids and biota; complexation with humic substances most important transport process
Bioaccumulation	short-term process, is readily metabolized	short-term process, is readily metabolized	short-term process, is readily metabolized	variety of organisms accumulate phthalates (lipophilic)	variety of organisms accumulate phthalates (lipophilic)	variety of organisms accumulate phthalates (lipophilic)	variety of organisms accumulate phthalates (lipophilic)
Biotransformation	readily metabolized by organisms; biodegradation probably ultimate fate mechanism	readily metabolized by organisms; biodegradation probably ultimate fate mechanism	readily metabolized by organisms; biodegradation probably ultimate fate mechanism	can be metabolized	can be metabolized	can be metabolized	can be metabolized

[a] Oxidation and hydrolysis are not important fate mechanisms for any of these compounds.

Source: Callahan, et al. 1979.

Table 35. Importance of Environmental Processes for the Aquatic Fates of Various Phenols and Pyrene

Environmental Process[a]	Phenol	Pentachlorophenol (PCP)	2,4,6-Trichlorophenol	2,4-Dimethyl Phenol (2,4-Xylene)	Pyrene
Photolysis	photooxidation may be important in degradation process in aerated, clear, surface waters	reported to occur in natural waters; important near water surface	reported, but importance is uncertain	may be important degradation process in clear aerated surface waters	dissolved portion may undergo rapid photolysis
Oxidation	metal-catalyzed oxidation may be important in aerated surface waters	not important	not important	metal-catalyzed oxidation may be important in aerated surface waters	not important
Volatilization	possibility of some phenol passing into the atmosphere	not important	not important	not important	not as important as adsorption
Sorption	not important	sorbed by organic litter in soil and sediments	potentially important for organic material, not important for clays	not important	adsorption onto suspended solids important; movement by suspended solids important
Bioaccumulation	not important	bioaccumulated in numerous aquatic organisms	not important	not important	short-term process not significant; metabolized over long term
Biotransformation	not important	can be metabolized to other phenol forms	reported in soil and sewage sludge; uncertain for natural surface waters	inconclusive information	readily metabolized; biodegradation probably ultimate fate process

[a] Hydrolysis is not an important fate mechanism for any of these compounds.

Source: Callahan, et al. 1979.

larger detention pond. The project located in Glen Ellyn (west of Chicago) monitored a small lake—the largest detention pond monitored during the NURP program. Ann Arbor, Michigan, monitoring included three detention ponds; Long Island, New York, studied one pond; the Washington, D.C. project included one pond. About 150 storm events were completely monitored at these ponds, and long-term performances ranged from negative removals for the smallest up-sized pipe installation to more than 90% removal of suspended solids at the largest wet ponds. The best ponds reported BOD_5 and COD removals of about 70%, nutrient removals of about 60 to 70%, and heavy metal removals of about 60 to 95%.

The Lansing NURP project monitored a wet detention pond (Luzkow et al. 1981). The monitored pond was located on a golf course that received urban runoff from an adjacent residential and commercial area. Suspended solids removals were about 70% for moderate rains (0.4- to 1-inch rains), while phosphorus removals were usually greater than 50%. Total Kjeldahl nitrogen removals ranged from about 30 to 50%.

Two wet detention ponds near Toronto, Ontario, were monitored from 1977 through 1979 (Brydges and Robinson 1980). Lake Aquitaine is 4.7 acres in size and receives runoff from a 107-acre urban watershed. Observed pollutant reductions were about 70 to 90% for suspended solids, 25 to 60% for nitrogen, and about 80% for phosphorus. The much smaller Lake Wabukayne (2 acres) received runoff from a much larger urban area (466 acres). Lake Wabukayne experienced much smaller pollutant reductions: about 30% for suspended solids, less than 25% for nitrogen, and 10 to 30% for phosphorus.

Oliver and Grigoropoulos (1981) monitored a small lake detention facility in Rolla, Missouri. Suspended solids yield reductions averaged about 88%, with 54 and 60% yield reductions for COD and total phosphorus. Organic nitrogen yields were reduced by about 22%.

Gietz (1983) studied a 3.3-acre wet detention pond serving a 150-acre urban watershed near Ottawa, Ontario. He compared batch operation (which retains water in the pond without discharge as long as possible) with normal, continuous operation (which has variable but continuous discharges). Batch operation of the pond resulted in substantial pollutant control improvements for particulate solids, bacteria, phosphorus, and nitrate nitrogen. Continuous operation gave slightly better performance for BOD_5 and organic nitrogen. Particulate solid reductions were about 80 to 95%, BOD_5 reductions were about 35 to 45%, bacteria was reduced by about 50 to 95%, phosphorus by about 70 to 85%, and organic nitrogen by about 45 to 50%.

Yousef et al. (1986) reported long-term nutrient removal information for a wet detention pond in Florida having substantial algal and rooted aquatic plant growths. He found 80 to 90% removals of soluble nutrients due to plant uptake. Particulate nutrient removals, however, were quite poor (about 10%).

Catchbasin, Sewerage, and Street Cleaning

The mobility of catchbasin sediments was investigated by Pitt (1979) during a research project sponsored by the U.S. EPA's Storm and Combined Sewer Section. This project used particulate fluorescent tracers mixed with catchbasin sediment. It was concluded that the amount of catchbasin and sewerage sediment was very large in comparison with storm runoff yields, but was not very mobile. Cleaning the material from catchbasins would reduce the potential of very large discharges during rare scouring rains.

Further research was conducted in Bellevue, Washington (Pitt 1984) to investigate the accumulation rate of sediment in storm sewerage and the effects of sewerage cleaning on runoff discharges. The main source of the sediment in the catchbasins and the sewerage was found to be the street surfaces. The catchbasin and sewerage sediment consisted of the largest particles that were washed from the streets. Smaller particles that had washed from the streets during rains had proceeded into receiving waters, leaving behind the larger particles. A few unusual locations were dominated by erosion sediment originating from steep hillsides adjacent to the storm sewer inlets.

Catchbasin sump particulates can be conveniently removed to eliminate this potential source of urban runoff pollutants and to enable the most effective capture of the larger particulates in the storm runoff. Cleaning catchbasins twice a year was found to be most effective. This cleaning schedule was found to reduce the total solids and lead urban runoff yields by between 10 and 25%, and COD, total Kjeldahl nitrogen, total phosphorus, and zinc by between 5 and 10% (Pitt 1984; Pitt and Shawley 1982).

Street cleaning effectiveness has been monitored at many locations and shows mixed results in removing toxicants from stormwater (Pitt 1979, Pitt and Shawley 1982, Bannerman, et al. 1983, and Pitt 1984, for example). Street cleaning has been shown to be very effective in removing the largest particulates on streets (especially those greater than about 200 μm). Unfortunately, street cleaning generally removes only a very small fraction of the small particles that are readily washed off streets by rains (Pitt 1987). Many of the street cleaning demonstration projects monitored a wide variety of available street cleaning equipment types, including mechanical broom sweepers, vacuum cleaners, and regenerative-air cleaners. Pitt (1984) also monitored a special prototype regenerative-air cleaner specifically modified to increase the removal of small particles. Many types of cleaner operations were investigated in some of these projects, including multiple passes using broom sweepers followed by vacuum cleaners, cleaning frequencies as often as two passes per day, full-street-width street cleaning, etc. Many demonstration projects only monitored relatively small changes in the street cleaning programs, however, and dramatic results could not have been expected.

In arid areas of the west, Pitt (1979) and Pitt and Shawley (1981) found that street cleaning could be beneficial in improving the stormwater quality associated with early fall rains following long dry summers. The dry summers allowed very large street dirt loadings to accumulate (if no street cleaning was used). If frequent street cleaning was used in the late summer (about weekly cleaning during September and October, for example), moderate heavy metal removals (25 to 50%) from the stormwater are likely. In most areas of the U.S., frequent rains would be more successful in keeping the streets clean than intensive street cleaning (Pitt 1984), resulting in very limited benefits.

Street runoff has also been overemphasized as a source of runoff pollutants for many areas. In most locations, streets contribute only a small portion of the total annual runoff loading, even though they are very important pollutant sources for the smallest rains (Pitt 1987). Therefore, even absolute cleanliness of streets would only result in limited overall stormwater quality improvement. In general, recommended street cleaning programs (cleaning about every three months in residential and commercial areas, and monthly in industrial areas, intensive spring cleaning after snowmelt in northern areas, rapid leaf removal in the fall, and intensive late summer cleaning in the arid west) using any type of street cleaning equipment available would result in optimal, but limited, stormwater quality improvements. Other stormwater control options are usually found to be more cost-effective in removing pollutants from stormwater than street cleaning (Pitt 1986).

Fate Mechanisms in Sedimentation Devices

The major fate mechanism in wet detention ponds and in smaller sumps, such as catchbasins, is sedimentation. Pollutants mostly associated with particulate matter will be much better removed than pollutants mostly in filterable forms. Unfortunately, sedimentation can result in the development of polluted sediments. These sediments can be anaerobic, causing various chemical and biochemical transformations. Resulting toxic chemical releases from heavily polluted sediments, plus the potential problems associated with the disposal of toxicant-contaminated dredging spoils during required maintenance, can present residual management problems.

Other important fate mechanisms available in wet detention ponds, but which are probably not important in small sump devices, include volatilization and photolysis. Biodegradation, biotransformation, and bioaccumulation (into plants and animals) may also occur in ponds. Most wet detention ponds are completely flushed by moderate rains (probably every several weeks), depending on their design. Much of the runoff during moderate and large rains passes through the ponds during several hours during and immediately after rains. Sediments may reside in ponds for several to many years. Therefore, the time available for these other removal or transformation processes can vary greatly for detention ponds. The residence time in small sedimentation devices is just a few minutes, and significant biological activity may not be present, except in the anaerobic sediments in catchbasin sumps and in the culverts of sewerage.

Most sedimentation devices (especially ponds) are designed to provide effective sedimentation and sufficient sacrificial storage for the long-term maintenance of accumulated sediment. The removal of many toxicants by other processes can possibly be increased by aerating the water and increasing the associated oxygen content and biological activity in ponds (Pitt, et al. 1995).

LOCAL PRETREATMENT OPTIONS BEFORE SOURCE AREA STORMWATER INFILTRATION

Biofiltration Devices

General Infiltration

All infiltration devices redirect surface runoff waters to the groundwater. They are recharge devices that can be used at many local areas in a watershed area. They must be carefully designed, especially using appropriate pretreatment as needed, to enable long-term operation and to protect groundwater quality.

Upland infiltration devices (such as infiltration trenches, porous pavements, percolation ponds, and grass roadside drainage swales) are located at urban source areas. Infiltration (percolation) ponds are usually located at stormwater outfalls, or at large paved areas. These ponds, along with perforated storm sewerage, can infiltrate flows and pollutants from all upland sources combined.

Several Nationwide Urban Runoff Program projects investigated infiltration devices (EPA 1983). They found that infiltration devices can safely deliver large fractions of the surface flows to groundwater, if carefully designed and located. Local conditions that can make localized stormwater infiltration inappropriate include steep slopes, slowly percolating soils, high groundwater, and nearby important groundwater uses.

The Lake Tahoe (California/Nevada) Regional Planning Agency has developed a set of design guidelines for infiltration devices in an area that has severe winters (Lake Tahoe 1978). They recommend the use of infiltration trenches to collect and infiltrate runoff from impervious surfaces, such as driveways, roofs, and parking lots. The Ontario Ministry of the Environment (1984) also included infiltration devices in its general stormwater management plan. The states of Florida, Maryland, and Delaware all extensively use infiltration as an important stormwater management tool and to recharge shallow groundwaters adversely affected by development. However, serious operational problems are very common with infiltration trenches and percolation ponds due to poor maintenance, poor construction practices, and poor placement (Lindsey, et al. 1992).

The Long Island, New York, and metropolitan Washington, D.C. NURP projects (EPA 1983) examined the performance of several types of infiltration devices. The Long Island project studied a series of interconnected percolating catchbasins, which were found to recharge more than 99% of the stormwater discharges. The Washington, D.C. study found that porous pavement recharged 85 to 95% of the pavement runoff flows, while an infiltration trench recharged about 50% of the surface flows. The EPA (1983) concluded that, with a reasonable degree of site-specific design considerations to compensate for soil characteristics, infiltration devices can be very effective in controlling urban runoff through recharging groundwaters.

Grass Filter Strips

Grass filter strips may be quite effective in removing pollutants from overland flows. The filtering effects of grasses, along with increased infiltration/recharge, reduce the pollutant load from urban landscaped areas. Filter strips are extensively used in contour strip cropping systems in agricultural areas to reduce erosion yields associated with grain crop production. Grass filters can be used at urban runoff source areas to reduce the particulate pollutant yields to the storm drainage system. Specific situations may include directing roof runoff to grassed areas instead of pavement, planting grass between eroding slopes and the storm drainage system, and planting grass between paved or unpaved parking or storage areas and the drainage system.

Novotny and Chesters (1981) reviewed several publications describing research on the effectiveness of grass filter strips. Grass filtering occurs during shallow flows, requiring the depth of flow to be less than the vegetation height. The critical length of the grass filter is defined as the minimum water flow length within which almost 100% of the particles of concern are removed. This length (and removal efficiency) varies for different particle sizes, grass density, flow depth, and flow velocity. For Bermuda grass, the critical length was found to be about 10 ft for sand, about 50 ft for silt, and about 400 ft for clay (Wilson 1967).

Grass Swales

Grass swale drainages are a type of infiltration device and can be used in place of concrete curb and gutter drainages in most land uses, except possibly strip commercial and high-density residential areas. Grass swales allow the recharge of significant amounts of surface flows.

Several large-scale urban runoff monitoring programs have included test sites that were drained by grass swales. Bannerman, et al. (1979), as part of the International Joint Commission (IJC) monitoring program to characterize urban runoff inputs to the Great Lakes, monitored a residential area served by swales and a similar residential area served by concrete curb and gutters in the Menominee River watershed in the Milwaukee area. This monitoring program included extensive flow and pollutant concentration measurements during a variety of rains. They found that the swale-drained area, even though it had soils characterized as poorly drained, had significantly lower surface flows (up to 95% lower) compared to the curb and gutter area.

The ability of grass swales to infiltrate source area sheetflows was also monitored in Durham, New Hampshire (EPA 1983). A special swale was constructed to treat runoff from a commercial parking lot. Flow measurements were not available to directly measure infiltration, but significant pollutant concentration reductions were found, apparently due to filtration. Soluble and particulate heavy metal (copper, lead, zinc, and cadmium) concentrations were reduced by about 50%. COD, nitrate nitrogen, and ammonia nitrogen concentrations were reduced by about 25%, while no significant concentration reductions were found for organic nitrogen, phosphorus, and bacteria.

Wang, et al. (1980) monitored the effectiveness of grass swales at several freeway sites in the state of Washington. Lead was more consistently and effectively trapped in the swale soils than the other metals, possibly because of its greater association with particulates in the runoff. Particulate filtering was therefore an important process during these tests. Lead concentrations were typically reduced by 80% or more, while copper was reduced by about 60%, and zinc by about 70%. Because of the particulate filtering action, they concluded that it may be necessary to remove the contaminated soils and replant the grass periodically to prevent dislodging the deposited polluted sediment. Part of the swales monitored by Wang, et al. (1980) were bare-earth lined. Pollutant concentrations were not found to be effectively reduced in these sections, and the earth lining was not contaminated.

A project to specifically study the effects of grass swale drainages was also conducted in Brevard County, Florida, by Kercher, et al. (1983). Two adjacent low-density residential areas, with about 14 acres and 50 homes each, were selected for study. One area had conventional concrete curbs and gutters, while the other had grass swales for roadside drainage. The two areas had very similar characteristics (soils, percentage imperviousness, slopes, vegetation, etc.). Thirteen storm events were monitored in the areas for flow and several selected pollutants. The curb and gutter area produced runoff flows during all 13 events, while the grass swale area produced runoff during only three events. The grass swale system also cost about one-half as much to construct as the curb and gutter system.

In another large-scale urban runoff monitoring project, Pitt and McLean (1985) monitored a residential area in Toronto that was served about evenly by swales and concrete curbs and gutters. The pollutant concentrations in both types of drainage systems were similar, but the area had annual flows about 25% less than if the area were served solely by curbs and gutters. For small but frequent rains (less than about 0.5 in.), very little surface runoff was observed in the swale sections, with almost all of the flows being infiltrated to the groundwater.

Porous Pavements

Porous pavement is a "hard" surface that can support a certain amount of activity, while still allowing water to pass through to recharge the underlying groundwaters. Porous pavement is generally used in areas of low traffic, such as service roads, storage areas, and parking lots. Several different types of porous pavement exist. Open mixes of asphalt appear similar to regular asphalt, but use only a specific size range of rocks in the hot mix. The porosity of the finished asphalt is much higher than regular asphalt, if properly designed and constructed. Concrete grids have open holes up to several inches wide filled with sand or gravel. It is possible to plant grass in the holes, if traffic is very light and if light and moisture conditions are adequate. They can be designed to recharge all of the runoff water from paved areas. The percolation rate of the pavement base is usually the limiting condition in porous pavement installations (Cedergren 1974).

Porous pavements provide some water quality treatment for the infiltrating water (William James, University of Guelph, personal communication), and allow the water to pass through soil before reaching the groundwater for further treatment. However, the organic content of the soil and associated sorption capacity of the soils may be limited below porous pavements because of the pavement construction operations.

Porous pavements may be effectively used in areas having soils with adequate percolation characteristics, if carefully designed and maintained. The percolation requirements for porous pavements are not as demanding as they are for other infiltration devices, unless runoff from other areas is directed toward the paved area. The percolation of the soils underlying the porous pavement installation need only exceed the rain intensity directly. In most cases, several inches of storage is available in the asphalt base to absorb short periods of very high rain intensities. Diniz (1980) states that the entire area contributing to the porous pavement can be removed from the surface hydrologic regime (and therefore be used for groundwater recharge).

Gburek and Urban (1983) studied a porous pavement parking lot in Pennsylvania. They found that percolation below the pavement occurred soon after the start of rain. For small rains (less than 0.25 in.), no percolation under the pavement was observed, with all of the rain being contained in the pavement base. Percolation during large rains was equal to about 70 to 90% of the rainfall. The differences between the rain amounts and the observed percolation quantities were caused by flash evaporation (not estimated) and storage in the asphalt base material (likely most important).

Goforth et al. (1983 and 1984) evaluated a porous pavement parking lot in Austin, Texas, over several years and under heavy traffic conditions. Infiltration rates through the pavement averaged about 1800 in./hr, while the 2-in. pavement base had an infiltration rate of about 70,000 in./hr. Day (1980) conducted a series of laboratory tests using several different types of concrete grid pavements. The geometry of the grid was more important than the percentage of open space in determining the ability of the grid to absorb and detain rainwater. The runoff coefficients from the grids ranged from 0.06 to 0.26 (resulting in recharge rates from about 75 to 95%) depending on the rain intensity, ground slope, and subsoil type.

Fate Mechanisms in Biofiltration Devices

Sorption of pollutants to soils is probably the most important fate mechanism of toxicants in biofiltration devices. Many of the devices also use sedimentation and filtration to remove the particulate forms of the pollutants from the water. Incorporation of the pollutants onto soil, with subsequent biodegradation and minimal leaching to the groundwater, is desired. Volatilization, photolysis, biotransformation, and bioconcentration may also be important in grass filter strips and grass swales. Underground French drains and porous pavements offer little biological activity to reduce toxicants.

REFERENCES

Alhajjar, B.J., G.V. Simsiman and G. Chesters. "Fate and Transport of Alachlor, Metolachlor and Atrazine in Large Columns." *Water Science and Technology*. Volume 22, number 6, pp. 87–94. 1990.

Amoros, I., J.L. Alonso and I. Peris. "Study of Microbial Quality and Toxicity of Effluents from Two Treatment Plants Used for Irrigation." *Water Science and Technology*. Volume 21, number 3, pp. 243–246. 1989.

APWA (American Public Works Association). *Water Pollution Aspects of Urban Runoff*. Water Pollution Control Research Series WP-20-15, Federal Water Pollution Control Administration, January 1969.

Armstrong, David E. and Reynaldo Llena. *Stormwater Infiltration: Potential for Pollutant Removal*. Report prepared for the Wisconsin Department of Natural Resources (Madison) and the U.S. Environmental Protection Agency, Chicago. February 1992.

Aronson, D.A. and G.E. Seaburn. *Appraisal of Operating Efficiency of Recharge Basins on Long Island, New York*. U.S. Geological Survey Water-Supply Paper 2001-D. USGS, Washington, D.C., 1974.

AWWA (American Water Works Association). "Fertilizer Contaminates Nebraska Groundwater." *AWWA Mainstream*. Volume 34, number 4, p. 6. April 1990.

Bannerman, R., K. Baun, M. Bohn, P.E. Hughes and D.A. Graczyk. *Evaluation of Urban Nonpoint Source Pollution Management in Milwaukee County, Wisconsin*, Vol. I. Grant No. P005432-01-5, PB 84-114164. US Environmental Protection Agency, Water Planning Division, November 1983.

Bannerman, R., J.G. Konrad and D. Becker. *The IJC Menominee River Watershed Study*. EPA-905/4-79-029. US Environmental Protection Agency. Chicago, Illinois. 1979.

Bao-rui, Yan. "Investigation into Mechanisms of Microbial Effects on Iron and Manganese Transformations in Artificially Recharged Groundwater." *Water Science and Technology*. Volume 20, number 3, pp. 47–53. 1988.

Basu, D. and S.J. Bosch. Bis (2-Chloroisopropyl) Ether Reportable Quantity (RQ) Ranking Based on Chronic Toxicity. Environmental Criteria and Assessment Office. U.S. Environmental Protection Agency, Contract No. 68-03-3112. Cincinnati, Ohio. July 1982.

Berg, G. Editor. *Transmission of Viruses by the Water Route*. Interscience Publishers, New York. 1965.

Boggess, D.H. *Effects of a Landfill On Ground-Water Quality*. United States Department of the Interior Geological Survey, Open File Report 75-594. Prepared in cooperation with the City of Fort Myers. U.S. Government Printing Office, Washington, D.C. 1975.

Bouwer, Edward J., Perry L. McCarty, Herman Bouwer and Robert C. Rice. "Organic Contaminant Behavior During Rapid Infiltration of Secondary Wastewater at the Phoenix 23rd Avenue Project." *Water Resources.* Volume 18, number 4, pp. 463–472. 1984.

Bouwer, Herman. "Renovation of Wastewater with Rapid-Infiltration Land Treatment Systems." In: *Artificial Recharge of Groundwater.* Edited by Takashi Asano. Butterworth Publishers, Boston, pp. 249–282. 1985.

Bouwer, Herman. "Effect of Irrigated Agriculture on Groundwater." *Journal of Irrigation and Drainage Engineering, ASCE.* Volume 113, number 1, pp. 4–15. February 1987.

Bouwer, Herman and Emannuel Idelovitch. "Quality Requirements for Irrigation with Sewage Water." *Journal of Irrigation and Drainage.* Volume 113, number 4, pp. 516–535. November 1987.

Bouwer, Herman. "Agricultural Contamination: Problems and Solutions." *Water Environment and Technology.* Volume 1, number 2, pp. 292–297. October 1989.

Bowman, D.C., J.L. Paul, W.B. Davis and S.H. Nelson. "Reducing Ammonia Volatilization from Kentucky Bluegrass by Irrigation." *Horticulture Science.* Vol. 22, pp. 84–87. 1987.

Brown, David P. *Effects of Effluent Spray Irrigation on Groundwater at a Test Site near Tarpon Springs, Florida.* United States Department of the Interior Geological Survey Open File Report 81-1197. Prepared in cooperation with Pinellas County, Florida. USGS, Denver, Colorado. 1982.

Brydges, T. and G. Robinson. "Two Examples of Urban Stormwater Impoundment for Aesthetics and for Protection of Receiving Waters." In: *Restoration of Lakes and Inland Waters: International Symposium on Inland Waters and Lake Restoration.* Portland, Maine, September 1980. Proceedings published by the U.S. Environmental Protection Agency, Washington, D.C. 1986.

Butler, Kent S. "Urban Growth Management and Groundwater Protection: Austin, Texas." *Planning for Groundwater Protection.* Academic Press, Inc. New York. pp. 261–287. 1987.

Callahan, M.A., M.W. Slimak, N.W. Gabel, I.P. May, C.F. Fowler, J.R. Freed, P. Jennings, R.L. Durfee, F.C. Whitmore, B. Maestri, W.R. Mabey, B.R. Holt and C. Gould. *Water Related Environmental Fates of 129 Priority Pollutants.* U.S. Environmental Protection Agency, Monitoring and Data Support Division, EPA-4-79-029a and b. Washington D.C., 1979.

Cedergren, H.R. *Drainage of Highway and Airfield Pavements.* John Wiley & Sons, New York, 1974.

Chang, A.C., A.L. Page, P.F. Pratt and J.E. Warneke. "Leaching of Nitrate from Freely Drained-Irrigated Fields Treated with Municipal Sludge." *Planning Now for Irrigation and Drainage in the 21st Century: Proceedings of a Conference Sponsored by the Irrigation and Drainage Division of the American Society of Civil Engineers.* Lincoln, Nebraska, pp. 455–467. ASCE, New York. July 18–21, 1988.

Chase, William L. "Reclaiming Wastewater in Phoenix, Arizona." *Irrigation Systems for the 21st Century - Proceedings of a Conference Sponsored by the Irrigation and Drainage Division of the American Society of Civil Engineers.* Portland, Oregon, July 28–30, 1987. ASCE, New York. pp. 336–343. 1987.

Close, M.E. "Effects of Irrigation on Water Quality of a Shallow Unconfined Aquifer." *Water Resources Bulletin.* Volume 23, number 5, pp. 793–802. October 1987.

Craun, Gunther F. "Waterborne Disease—A Status Report Emphasizing Outbreaks in Ground-Water Systems." *Groundwater.* Volume 17, number 2, pp. 183–191. March–April 1979.

Crites, Ronald W. "Micropollutant Removal in Rapid Infiltration." In: *Artificial Recharge of Groundwater.* Edited by Takashi Asano. Butterworth Publishers, Boston, pp. 579–608. 1985.

Dalrymple, R.J., S.L. Hodd and D.C. Morin. *Physical and Settling Characteristics of Particulates in Storm and Sanitary Wastewaters.* EPA-670/2-75-011. U.S. Environmental Protection Agency, Cincinnati, Ohio, 1975.

Day, G.E. "Investigation of Concrete Grid Pavements." In *Proceedings - National Conference on Urban Erosion and Sediment Control: Institutions and Technology*, EPA-905/9-80-002, U.S. Environmental Protection Agency, Chicago, Ill., January 1980.

Deason, Jonathan P. "Selenium: It's Not Just in California." *Irrigation Systems for the 21st Century—Proceedings of a Conference.* Sponsored by the Irrigation and Drainage Division of the American Society of Civil Engineers, New York. pp. 475–482. 1987.

Deason, Jonathan P. "Irrigation-Induced Contamination: How Real a Problem?" *Journal of Irrigation and Drainage Engineering, ASCE.* Volume 115, number 1, pp. 9–20. February 1989.

DeBoer, Jon G. "Wastewater Reuse: A Resource or a Nuisance?" *Journal of the American Water Works Association.* Volume 75, pp. 348–356. July 1983.

Diniz, E.V. *Porous Pavement; Phase I, Design and Operational Criteria.* EPA-600/2-80-135. U.S. Environmental Protection Agency, Cincinnati, Ohio, August 1980.

Domagalski, Joseph L. and Neil M. Dubrovsky. "Pesticide Residues in Groundwater of the San Joaquin Valley, California." *Journal of Hydrology.* Volume 130, number 1–4, pp. 299–338. January 1992.

Durum, W.H. Occurrence of Some Trace Metals in Surface Waters and Groundwaters. In Proceeding of the Sixteenth Water Quality Conference, Am. Water Works Assoc., et al. Univ. of Illinois Bull., 71(108), Urbana, Illinois. 1974.

Ehrlich, Garry G., Henry F.H. Ku, John Vecchioli and Theodore A. Ehlke. *Microbiological Effects of Recharging the Magothy Aquifer, Bay Park, New York, with Tertiary-Treated Sewage.* Geological Survey Professional Paper 751-E. Prepared in cooperation with the Nassau County Department of Public Works. USGS, Washington, D.C. 1979a.

Ehrlich, G.G., E.M. Godsy, C.A. Pascale and John Vecchioli. "Chemical Changes in an Industrial Waste Liquid during Post-Injection Movement in a Limestone Aquifer, Pensacola, Florida." *Groundwater.* Volume 17, number 6, November–December, pp. 562–573. 1979b.

Environment Canada. Rideau River Water Quality and Stormwater Monitoring Study. MS Report OR-29, Ontario Ministry of the Environment. February 1980.

Environment Canada/Agriculture Canada. *Pesticide Registrant Survey; 1986 Report.* Prepared by the Commercial Chemicals Branch. Conservation and Protection. Environment Canada. January 1987.

EPA. *Areawide Assessment Manual.* Three Volumes. Municipal Envir. Research Lab., Cincinnati, Ohio, July 1976.

EPA. *Results of the Nationwide Urban Runoff Program.* Water Planning Division, PB 84-185552, Washington, D.C., December 1983.

EPA. *Quality Criteria for Water.* U.S. Environmental Protection Agency. EPA 440/5-86-001. Washington., D.C., May 1986.

EPA. *National Pesticide Survey: Summary Results of EPA's National Survey of Pesticides in Drinking Water Wells.* EPA Office of Pesticides and Toxic Substances. U.S. Government Printing Office, Washington, D.C. 1990.

EPA. *Manual: Guidelines for Water Reuse.* EPA Document No. EPA/625/R-92/004. U.S. Government Printing Office, Washington, D.C. September, 1992.

Ferguson, Bruce K. "Role of the Long-Term Water Balance in Management of Stormwater Infiltration." *Journal of Environmental Management.* Volume 30, number 3, pp. 221–233. April 1990.

Ferguson, R.B., D.E. Eisenbauer, T.L. Bockstadter, D.H. Krull and G. Buttermore. "Water and Nitrogen Management in Central Platte Valley of Nebraska." *Journal of Irrigation and Drainage.* Volume 116, number 4, pp. 557–565. July/August 1990.

Field, R., E.J. Struzeski, Jr., H.E. Masters and A.N. Tafuri. Water Pollution and Associated Effects from Street Salting. EPA-R2-73-257, U.S. Environmental Protection Agency, Cincinnati, Ohio. May 1973.

Field, R. and M.L. O'Shea. "The Detection of Pathogens in Storm-Generated Flows." *Water Environment Federation 65th Annual Conference and Exposition.* New Orleans. September, 1992.

Galvin, D.V. and R.K. Moore. *Toxicants in Urban Runoff. Toxicant Control Planning Section, Municipality of Metropolitan Seattle,* Contract # P-16101, U.S. Environmental Protection Agency, Lacy, Washington, December 1982.

Gburek, W.J. and J.B. Urban. "Storm Water Detention and Groundwater Recharge Using Porous Asphalt - Initial Results." *1983 International Symposium on Urban Hydrology, Hydraulics and Sediment Control,* University of Kentucky, Lexington, Kentucky, July 1983.

Gerba, Charles P. and Sagar M. Goyal. "Pathogen Removal from Wastewater during Groundwater Recharge." In: *Artificial Recharge of Groundwater.* Edited by Takashi Asano. Butterworth Publishers, Boston. pp. 283–317. 1988.

Gerba, Charles P. and Charles N. Haas. "Assessment of Risks Associated with Enteric Viruses in Contaminated Drinking Water." *Chemical and Biological Characterization of Sludges, Sediments, Drudge Spoils and Drilling Muds, ASTM STP 976.* American Society for Testing and Materials, Philadelphia, Pennsylvania. pp. 489–494. 1988.

Geldreich, E.E. and B.A. Kenner. Concepts of Fecal Streptococci in Stream Pollution. Journal WPCF, Vol. 41, No. 8, pp. R336-R352. Aug. 1969.

German, Edward R. *Quantity and Quality of Stormwater Runoff Recharged to the Floridan Aquifer System Through Two Drainage Wells in the Orlando, Florida Area.* U.S. Geological Survey - Water Supply Paper 2344. Prepared in cooperation with the Florida Department of Environmental Regulation. USGS, Denver, Colorado. 1989.

Gietz, R.J. *Urban Runoff Treatment in the Kennedy-Burnett Settling Pond.* For the Rideau River Stormwater Management Study, Pollution Control Division, Works Department, Regional Municipality of Ottawa-Carleton, Ottawa, Ontario, March 1983.

Goforth, G.F., E.V. Diniz and J.B. Rauhut. *Stormwater Hydrological Characteristics of Porous and Conventional Paving Systems.* EPA-600/2-83-106, U.S. Environmental Protection Agency, Cincinnati, Ohio, February 1984 (also dated October 1983).

Gold, A.J. and P.M. Groffman. "Leaching of Agrichemicals from Suburban Areas." In: *Pesticides in Urban Environments—Fate and Significance.* Racke, K.D. and A.R. Leslie, Editors. ACS Symposium Series No. 522. American Chemical Society, Washington, D.C. 1993.

Goldshmid, J. "Water-Quality Aspects of Ground-Water Recharge in Israel." *Journal of the American Water Works Association.* Volume 66, number 3, pp. 163–166. March 1974.

Goolsby, Donald A. "Geochemical Effects and Movement of Injected Industrial Waste in a Limestone Aquifer." *Underground Waste Management and Environmental Implications (Reprint).* The American Association of Petroleum Geologists, Memoir No. 18. 1972.

Gore & Storrie Ltd./Proctor & Redfern Ltd. Executive Summary Report on Rideau River Stormwater Management Study, Phase 1, Rideau River Stormwater Management Study, Ottawa and the Ontario Ministry of the Environment, Kingston, Ontario. 1981.

Greene, Gerald E. "Ozone Disinfection and Treatment of Urban Storm Drain Dry-Weather Flows: A Pilot Treatment Plant - Demonstration Project on the Kenter Canyon Storm Drain System in Santa Monica." *Santa Monica Bay Restoration Project.* Office of the City Engineer, Santa Monica, California. June 1992.

Hampson, Paul S. *Effects of Detention on Water Quality of Two Stormwater Detention Ponds Receiving Highway Surface Runoff in Jacksonville, Florida.* U.S. Geological Survey Water-Resources Investigations Report 86-4151. Prepared in cooperation with the Florida Department of Transportation. USGS, Denver, Colorado. 1986.

Harper, Harvey H. *Effects of Stormwater Management Systems on Groundwater Quality.* Final Report for DER Project WM190. Florida Department of Environmental Regulation. September 1988.

Harris, B. Fluoranthene: Reportable Quantity (RQ) Ranking Based on Chronic Toxicity. U.S. Environmental Protection Agency. Contract No. 68-03-3112. Cincinnati, Ohio, July 1982.

HEW (U.S. Department of Health, Education, and Welfare). Bioassay of Technical Grade Bis (2-Chloro-1-Methylethyl) Ether for Possible Carcinogenicity. CAS No. 108-60-1, NCI-CG-TR-191. Bethesda, Maryland, 1979.

Hickey, John J. and John Vecchioli. *Subsurface Injection of Liquid Waste with Emphasis on Injection Practices in Florida.* U.S. Geological Survey Water-Supply Paper 2281. USGS, Denver, Colorado. 1986.

Higgins, Andrew J. "Impacts on Groundwater due to Land Application of Sewage Sludge." *Water Resources Bulletin.* Volume 20, number 3, pp. 425–434. June 1984.

Horsley, Scott W. and John A. Moser. "Monitoring Groundwater for Pesticides at a Golf Course—A Case Study on Cape Cod, Massachusetts." *Groundwater Monitoring Review.* Volume 10, pp. 101–108. Winter 1990.

Howard, P.H. Handbook of Environmental Fate and Exposure Data for Organic Chemicals, Volume 1, Large Production and Priority Pollutants. Lewis Publishers. Chelsea, Michigan. 1989.

IARC. Chemicals and Industrial Processes Associated with Cancer in Humans. IARC Monographs on the Evaluation of the Carcinogenic Risk of Chemicals to Humans, Supplement 1. World Health Organization. Lyon, France, 1979.

Iowa DNR. *Pesticide and Synthetic Organic Compound Survey.* Iowa Department of Natural Resources. 1988.

Iowa DNR. *The Iowa State-Wide Rural Well-Water Survey Water Quality Data: Initial Analysis.* Iowa Department of Natural Resources. 1990.

Jansons, Janis, Lindsay W. Edmonds, Brent Speight and Marion R. Bucens. "Movement of Viruses after Artificial Recharge." *Water Research.* Volume 23, number 3, pp. 293–299. 1989a.

Jansons, Janis, Lindsay W. Edmonds, Brent Speight and Marion R. Bucens. "Survival of Viruses in Groundwater." *Water Research.* Volume 23, number 3, pp. 301–306. 1989b.

Johnson, C.R. "A New Look at Sewer Separation for CSO Control." *WPCF Specialty Conference Series: Control of Combined Sewer Overflows.* Boston. April 8–11, 1990.

Jury, W.A., W.F. Spencer and W.J. Farmer. "Model for Assessing Behavior of Pesticides and other Trace Organics using Benchmark Properties. I. Description of Model." *J. Environmental Quality.* No. 12. pp. 558–564. 1983.

Kercher, W.C., J.C. Landon and R. Massarelli. "Grassy Swales Prove Cost-Effective for Water Pollution Control." *Public Works*, April 1983.

Knisel, W.G. and R.A. Leonard. "Irrigation Impact on Groundwater: Model Study in Humid Region." *Journal of Irrigation and Drainage, ASCE.* Volume 115, number 5, pp. 823–838. October 1989.

Krawchuk, Bert P. and G.R. Barrie Webster. "Movement of Pesticides to Groundwater in an Irrigated Soil." *Water Pollution Research Journal of Canada.* Volume 22, pp. 129–146, number 1, 1987.

Ku, Henry F.H., John Vecchioli and Stephen E. Ragone. "Changes in Concentration of Certain Constituents of Treated Waste Water during Movement through the Magothy Aquifer, Bay Park, New York." *Journal Research U.S. Geological Survey.* Volume 3, number 1, pp. 89–92. January–February 1975.

Ku, Henry F.H. and Dale L. Simmons. *Effect of Urban Stormwater Runoff on Groundwater beneath Recharge Basins on Long Island, New York.* U.S. Geological Survey Water-Resources Investigations Report 85-4088. Prepared in cooperation with Long Island Regional Planning Board, Syosset, New York. USGS, Denver, Colorado. 1986.

Ku, Henry H.F., Nathan W. Hagelin and Herbert T. Buxton. "Effects of Urban Storm-Runoff Control on Ground-Water Recharge in Nassau County, New York." *Groundwater.* Volume 30, number 4, pp. 507–514. July–August 1992.

Lager, J.A. *Urban Stormwater Management and Technology: Update and User's Guide.* Rep No. EPA-600/8-77-014. U.S. Environmental Protection Agency. Cincinnati, Ohio. pp. 90–92. 1977.

Lake Tahoe Regional Planning Agency. *Lake Tahoe Basin Water Quality Management Plan, Volume II, Handbook of Best Management Practices.* Lake Tahoe, California, January 1978.

Lauer, D.A. "Vertical Distribution in Soil of Sprinkler-Applied Phosphorus." *Soil Science Society of America Journal.* Volume 52, number 3, pp. 862–868. May/June 1988a.

Lauer, D.A. "Vertical Distribution in Soil of Unincorporated Surface-Applied Phosphorus under Sprinkler Irrigation." *Soil Science Society of America Journal.* Volume 52, number 6, pp. 1685–1692. November/December 1988b.

Lee, Edwin W. "Chapter 21: Drainage Water Treatment and Disposal Options." In: *Agricultural Salinity Assessment and Management.* Edited by Kenneth K. Tanji. American Society of Civil Engineers, New York. pp. 450–468. 1990.

Lindsey, G., L. Roberts and W. Page. "Inspection and Maintenance of Infiltration Facilities." *Journal of Soil and Water Conservation.* Vol. 47, No. 6, pp. 481–486, November/December 1992.

Lloyd, J.W., D.N. Lerner, M.O. Rivett and M. Ford. "Quantity and Quality of Groundwater beneath an Industrial Conurbation—Birmingham, UK." In: *Proceedings of the Conference—Hydrological Processes and Water Management in Urban Areas.* Duisburg, Federal Republic of Germany. International Hydrological Programme, UNESCO, pp. 445–452. April 24–29, 1988.

Loague, K.M. and R.A. Freeze. "A Comparison of Rainfall-Runoff Modeling Techniques on Small Upland Catchments," *Water Resources Research*, Vol. 21, No. 2, pp 229–248, February 1985.

Luzkow, S.M., D.A. Scherger and J.A. Davis. "Effectiveness of Two In-Line Urban Stormwater Best Management Practices (BMP's)." *1981 International Symposium on Urban Hydrology, Hydraulics, and Sediment Control*, University of Kentucky, Lexington, Kentucky, July 1981.

Madison, F., J. Arts, S. Berkowitz, E. Salmon and B. Hagman. *Washington County Project.* EPA 905/9-80-003, U.S. Environmental Protection Agency, Chicago, IL, 1979.

Marton, J. and I. Mohler. "The Influence of Urbanization on the Quality of Groundwater." In: *Proceedings of the Conference—Hydrological Processes and Water Management in Urban Areas.* Duisburg, Federal Republic of Germany. International Hydrological Programme, UNESCO, pp. 453–460. April 24–29, 1988.

Marzouk, Yosef, Sagar M. Goyal and Charles P. Gerba. "Prevalence of Enteroviruses in Groundwater of Israel." *Groundwater.* Volume 17, number 5, pp. 487–491. September–October 1979.

McComas, S.R. and D.F. Sefton. "Comparison of Stormwater Runoff Impacts on Sedimentation and Sediment Trace Metals for Two Urban Impoundments." In: *Lake and Reservoir Management: Practical Applications*, proceedings of the Fourth Annual Conference and International Symposium, October, 1984, North American Lake Management Society, McAfee, New Jersey, 1985.

Merkel, B., J. Grossman and P. Udluft. "Effect of Urbanization on a Shallow Quaternary Aquifer." *Proceedings of the Conference—Hydrological Processes and Water Management in Urban Areas.* Duisburg, Federal Republic of Germany. International Hydrological Programme, UNESCO, pp. 461–468. April 24–29, 1988.

Moffa, P.E. *Control and Treatment of Combined Sewer Overflows.* Van Nostrand Reinhold. New York. 1989

Mossbarger, W.A. and R.W. Yost. "Effects of Irrigated Agriculture on Groundwater Quality in Corn Belt and Lake States." *Journal of Irrigation and Drainage Engineering, ASCE.* Volume 115, number 5, pp. 773–790. October 1989.

Natarajan, U. and R. Rajagopal. "Surveying The Situation." *Environmental Testing and Analysis*. pp. 40–50. Jan/Feb. 1993.

National Research Council. Ground Water Recharge Using Waters of Impaired Quality. National Academy Press, Washington, D.C., 1994.

Nellor, Margaret H., Rodger B. Baird and John R. Smyth. "Health Aspects of Groundwater Recharge." In: *Artificial Recharge of Groundwater*. Edited by Takashi Asano. Butterworth Publishers, Boston, pp. 329–356. 1985.

Nightingale, Harry I. and William C. Bianchi. "Ground-Water Chemical Quality Management by Artificial Recharge." *Groundwater*. Volume 15, number 1, pp. 15–22. January–February 1977a.

Nightingale, Harry I. and William C. Bianchi. "Ground-Water Turbidity Resulting from Artificial Recharge." *Groundwater*. Volume 15, number 2, pp. 146–152. March–April 1977b.

Nightingale, H.I., J.E. Ayars, R.L. McCormick and D.C. Cehrs. "Leaky Acres Recharge Facility: A Ten-Year Evaluation." *Water Resources Bulletin*. Volume 19, number 3, pp. 429–437. June 1983.

Nightingale, Harry I. "Water Quality beneath Urban Runoff Water Management Basins." *Water Resources Bulletin*. Volume 23, number 2, pp. 197–205. April 1987a.

Nightingale, Harry I. "Accumulation of As, Ni, Cu and Pb in Retention and Recharge Basin Soils from Urban Runoff." *Water Resources Bulletin*. Volume 23, number 4, pp. 663–672. August 1987b.

Norberg-King, Teresa J., Elizabeth J. Durhan, Gerald T. Ankley and Eric Robert. "Application of Toxicity Identification Evaluation Procedures to the Ambient Waters of the Colusa Basin Drain, California." *Environmental Toxicology and Chemistry*. Volume 10, pp. 891–900. 1991.

Novotny, V. and G. Chesters. *Handbook of Nonpoint Pollution*. Van Nostrand Reinhold, New York, 1981.

Oliver, L.J. and S.G. Grigoropoulis. "Control of Storm-Generated Pollution Using a Small Urban Lake." *Journal Water Pollution Control Federation*, Vol 53, No. 5, May 1981.

Olivieri, V.P., C.W. Kurse and K. Kawata. Selected Pathogenic Microorganisms Contributed from Urban Watersheds. In: Watershed Research in Eastern North America, Vol. II, D.L. Correll. NTIS No. PB-279 920/3SL. 1977a.

Olivieri, V.P., C.W. Kurse and K. Kawata. Microorganisms in Urban Stormwater, U.S. Environmental Protection Agency, EPA-600/2-77-087. July 1977b.

Ontario Ministry of the Environment. *Rideau River Stormwater Management Study*. Kingston, Ontario, 1983.

Ontario (Province of). *Urban Drainage Design Guidelines* (Draft). Urban Drainage Policy Implementation Committee, Technical Sub-Committee No. 3. March 1984.

Ontario Ministry of the Environment. *Humber River Water Quality Management Plan*, Toronto Area Watershed Management Strategy. Toronto, Ontario, 1986.

Peterson, David A. "Selenium in the Kendrick Reclamation Project, Wyoming." *Planning Now for Irrigation and Drainage in the 21st Century*. Proceedings of a conference sponsored by the Irrigation and Drainage Division of the American Society of Civil Engineers, Lincoln, Nebraska, pp. 678–685. ASCE, New York. July 18–21, 1988.

Petrovic, A. Martin. "The Fate of Nitrogenous Fertilizers Applied to Turfgrass." *Journal of Environmental Quality*. Volume 19, number 1, pp. 1–14. January–March 1990.

Phillips, G.R., and R.C. Russo. Metal Bioaccumulation in Fishes and Aquatic Invertebrates: A Literature Review. EPA-600-3-78-103, U.S. Environmental Protection Agency, Duluth, Minnesota. December 1978.

PHS (U.S. Public Health Service). Second Annual Report on Carcinogens. U.S. Department of Health and Human Services. Dec. 1981.

Pierce, Ronald C. and Michael P. Wong. "Pesticides in Agricultural Waters: The Role of Water Quality Guidelines." *Canadian Water Resources Journal*. Volume 13, number 3, pp. 33–49. July, 1988.

Pitt, R. and G. Amy. *Toxic Materials Analysis of Street Surface Contaminants*. EPA-R2-73-283, U.S. Environmental Protection Agency, Washington, D.C., August 1973.

Pitt, R. *Demonstration of Nonpoint Pollution Abatement Through Improved Street Cleaning Practices*. EPA-600/2-79-161, U.S. Environmental Protection Agency, Cincinnati, Ohio, August 1979.

Pitt, R. and G. Shawley. *A Demonstration of Non-Point Source Pollution Management on Castro Valley Creek*. Alameda County Flood Control and Water Conservation District (Hayward, CA) for the Nationwide Urban Runoff Program, U.S. Environmental Protection Agency, Water Planning Division, Washington, D.C., June 1982.

Pitt, R. *Urban Bacteria Sources and Control by Street Cleaning in the Lower Rideau River Watershed, Ottawa, Ontario*. Rideau River Stormwater Management Study, Ontario Ministry of the Environment, Ottawa, Ontario, 1983.

Pitt, R. *Characterization, Sources, and Control of Urban Runoff by Street and Sewerage Cleaning*. Contract No. R-80597012, U.S. Environmental Protection Agency, Office of Research and Development, Cincinnati, Ohio, 1984.

Pitt, R. and P. Bissonnette. *Bellevue Urban Runoff Program, Summary Report*. U.S. Environmental Protection Agency and the Storm and Surface Water Utility, Bellevue, Washington, 1984.

Pitt, R. "Runoff Controls in Wisconsin's Priority Watersheds." *Conference on Urban Runoff Impact and Quality Enhancement Technology. Henniker, NH*. Edited by B. Urbonas and L.A. Roesner, Proceedings published by the American Society of Civil Engineers. New York, NY. June 1986.

Pitt, R. and J. McLean. *Toronto Area Watershed Management Strategy Study: Humber River Pilot Watershed Project*. Ontario Ministry of the Environment, Toronto, Ontario, 1986.

Pitt, R. *Small Storm Urban Flow and Particulate Washoff Contributions to Outfall Discharges*. Ph.D. dissertation. Department of Civil and Environmental Engineering, University of Wisconsin–Madison. October 1987.

Pitt, R., and P. Barron. *Assessment of Urban and Industrial Stormwater Runoff Toxicity and the Evaluation/Development of Treatment for Runoff Toxicity Abatement—Phase I*. U.S. Environmental Protection Agency, Office of Research and Development, Edison, New Jersey, 1990.

Pitt, R.M. Lalor, R. Field, D.D. Adrian and D. Barbé. *Investigation of Inappropriate Pollutant Entries into Storm Drainage Systems*. EPA/600/R-92/238. U.S. Environmental Protection Agency. Office of Research and Development, Cincinnati, Ohio, 1993.

Pitt, R.E., R. Field, M. Lalor and M. Brown. "Urban Stormwater Toxic Pollutants: Assessment, Sources, and Treatability." *Water Environment Research*. Vol. 67, No. 3, pp. 260–275. May/June 1995.

Power, J.F. and J.S. Schepers. "Nitrate Contamination of Groundwater in North America." *Agriculture, Ecosystems and Environment*. Volume 26, pp. 165–187, number 3–4 1989.

Pruitt, Janet B., David A. Troutman and G.A. Irwin. *Reconnaissance of Selected Organic Contaminants in Effluent and Groundwater at Fifteen Municipal Wastewater Treatment Plants in Florida, 1983–84*. U.S. Geological Survey Water-Resources Investigation Report 85-4167. Prepared in cooperation with the Florida Department of Environmental Regulation. USGS. Denver, Colorado. 1985.

Qureshi, A.A. and B.J. Dutka. Microbiological Studies on the Quality of Urban Stormwater Runoff in Southern Ontario, Canada. Water Research, Vol. 13, pp. 977–985. 1979.

Racke, K.D. and A.R. Leslie, Editors. *Pesticides in Urban Environments—Fate and Significance*. ACS Symposium Series No. 522. American Chemical Society, Washington, D.C. 1993.

Ragone, Stephen E. *Geochemical Effects of Recharging the Magothy Aquifer, Bay Park, New York, with Tertiary-Treated Sewage.* Geological Survey Professional Paper 751-D. Prepared in cooperation with the Nassau County Department of Public Works. USGS, Washington, D.C. 1977.

Razack, M., C. Drogue and M'Baitelem. "Impact of an Urban Area on the Hydrochemistry of a Shallow Groundwater (Alluvial Reservoir) Town of Narbonne, France." *Proceedings of the Conference—Hydrological Processes and Water Management in Urban Areas.* International Hydrological Programme, UNESCO, pp. 487–494. Duisburg, Federal Republic of Germany, April 24–29, 1988.

Reichenbaugh, R.C. *Effects on Ground-Water Quality from Irrigating Pasture with Sewage Effluent Near Lakeland, Florida.* U.S. Geological Survey Water-Resources Investigations Report 76-108. Prepared in cooperation with Southwest Florida Water Management District. USGS, Denver, Colorado. April 1977.

Rice, R.C., D.B. Jaynes and R.S. Bowman. "Preferential Flow of Solutes and Herbicide under Irrigated Fields." *Transactions of the American Society of Agricultural Engineers.* Volume 34, number 3, pp. 914–918. May–June 1991.

Ritter, W.F., F.J. Humenik and R.W. Skaggs. "Irrigated Agriculture and Water Quality in East." *Journal of Irrigation and Drainage Engineering, ASCE.* Volume 115, number 5, pp. 807–821. October 1989.

Ritter, W.F., R.W. Scarborough and A.E.M. Chirnside. "Nitrate Leaching under Irrigation on Coastal Plain Soil." *Journal of Irrigation and Drainage Engineering, ASCE.* Volume 117, number 4, pp. 490–502. July/August 1991.

Robinson, J. Heyward and H. Stephen Snyder. "Golf Course Development Concerns in Coastal Zone Management." *Coastal Zone '91: Proceedings of the Seventh Symposium on Coastal and Ocean Management.* Long Beach, California, pp. 431–443. ASCE, New York. July 8–12, 1991.

Rolfe, G.L. and K.A. Reinhold. Vol. I: *Introduction and Summary. Environmental Contamination by Lead and Other Heavy Metals.* Institute for Environmental Studies, University of Illinois, Champaign-Urbana, Illinois, July 1977.

Sabatini, David A. and T. Al Austin. "Adsorption, Desorption and Transport of Pesticides in Groundwater: A Critical Review." *Planning Now for Irrigation and Drainage in the 21st Century.* Proceedings of a Conference Sponsored by the Irrigation and Drainage Division of the American Society of Civil Engineers, Lincoln, Nebraska, pp. 571–579. ASCE, New York. July 18–21, 1988.

Sabol, George V., Herman Bouwer and Peter J. Wierenga. "Irrigation Effects in Arizona and New Mexico." *Journal of Irrigation and Drainage Engineering, ASCE.* Volume 113, number 1, pp. 30–57. February 1987.

Saffigna, P.G. and D.R. Keeney. "Nitrate and Chloride in Groundwater under Irrigated Agriculture in Central Wisconsin." *Ground Water.* Vol. 15, no. 2. pp. 170–177. 1977.

Salo, John E., Doug Harrison and Elaine M. Archibald. "Removing Contaminants by Groundwater Recharge Basins." *Journal of the American Water Works Association.* Volume 78, number 9, pp. 76–81. September 1986.

Schiffer, Donna M. *Effects of Three Highway-Runoff Detention Methods on Water Quality of the Surficial Aquifer System in Central Florida.* U.S. Geological Survey Water-Resources Investigations Report 88-4170. Prepared in cooperation with the Florida Department of Transportation. USGS, Denver, Colorado. 1989.

Schillinger, J.E. and D.G. Stuart. *Quantification of Non-Point Water Pollutants from Logging, Cattle Grazing, Mining, and Subdivision Activities.* U.S. Environmental Protection Agency. PB 80-174063. 1978

Schmidt, Kenneth D. and Irving Sherman. "Effect of Irrigation on Groundwater Quality in California." *Journal of Irrigation and Drainage Engineering, ASCE.* Volume 113, number 1, pp. 16–29. February 1987.

Schneider, Brian J., Henry F.H. Ku and Edward T. Oaksford. *Hydrologic Effects of Artificial-Recharge Experiments with Reclaimed Water at East Meadow, Long Island, New York.* U.S. Geological Survey Water-Resources Investigations Report 85-4323. Prepared in cooperation with the Nassau County Department of Public Works, Syosset, New York. USGS, Denver, Colorado. 1987.

Seaburn, G.E. and D.A. Aronson. *Influence of Recharge Basins on the Hydrology of Nassau and Suffolk Counties, Long Island, New York.* U.S. Geological Survey Water-Supply Paper 2031. USGS, Washington, D.C. 1974.

Setmire, J.G. and W.L. Bradford. "Quality of Urban Runoff, Tecolote Creek Drainage Area, San Diego County, CA." U.S. Geological Survey. PB81-159451. Menlo Park, CA. 1980

Shirmohammadi, A. and W.G. Knisel. "Irrigated Agriculture and Water Quality in the South." *Journal of Irrigation and Drainage Engineering, ASCE.* Volume 115, number 5, pp. 791–806. October 1989.

Smith, A.E. "Herbicides and the Soil Environment in Canada." *Canadian Journal of Soil Science.* Volume 62. pp. 433–460. 1982.

Smith, Sheldon O. and Donald H. Myott. "Effects of Cesspool Discharge on Ground-Water Quality on Long Island, N.Y." *Journal of the American Water Works Association.* Volume 67, number 8, pp. 456–458. August 1975.

Smith, W.G. "Water Quality Enhancement through Stormwater Detention." *Proceedings of the Conference on Stormwater Detention Facilities, Planning, Design, Operation, and Maintenance*, Henniker, New Hampshire, Edited by W. DeGroot, published by the American Society of Civil Engineers, New York, August 1982.

Solomon, R.L. and D.F.S. Natusch. Vol. III: "Distribution and Characterization of Urban Districts." In: *Environmental Contamination by Lead and Other Heavy Metals,* G.L. Rolfe and K.G. Reinhold, Eds. Institute for Environmental Studies, Univ. of Illinois, Urbana-Champaign, Illinois. July 1977.

Spalding, Roy F. and Lisa A. Kitchen. "Nitrate in the Intermediate Vadose Zone Beneath Irrigated Cropland." *Groundwater Monitoring Review.* Volume 8, number 2, pp. 89–95. Spring 1988.

Squires, Rodney C. and William R. Johnston. "Selenium Removal—Can We Afford It? *Irrigation Systems for the 21st Century—Proceedings of a Conference.* Sponsored by the Irrigation and Drainage Division of the American Society of Civil Engineers. New York, New York, pp. 455–466. 1987.

Squires, Rodney C., G. Raymond Groves and William R. Johnston. "Economics of Selenium Removal from Drainage Water." *Journal of Irrigation and Drainage Engineering.* Volume 115, number 1, pp. 48–57. February 1989.

Steenhuis, Tammo, Robert Paulsen, Tom Richard, Ward Staubitz, Marc Andreini and Jan Surface. "Pesticide and Nitrate Movement under Conservation and Conventional Tilled Plots." *Planning Now for Irrigation and Drainage in the 21st Century.* Proceedings of a conference sponsored by the Irrigation and Drainage Division of the American Society of Civil Engineers, Lincoln, Nebraska, pp. 587–595. ASCE, New York. July 18–21, 1988.

Tetra Tech. *Detention and Retention Effects on Groundwater, A Literature Review.* Contract No. 68-C9-0013. U.S. Environmental Protection Agency. Watershed Management Unit. Water Division. Region V. Chicago, Ill. February 1991.

Tim, Udoyara S. and Saied Mostaghim. "Model for Predicting Virus Movement through Soil." *Groundwater.* Volume 29, number 2, pp. 251–259. March–April 1991.

Treweek, Gordon P. "Pretreatment Processes for Groundwater Recharge." In: *Artificial Recharge of Groundwater.* Edited by Takashi Asano. Butterworth Publishers, Boston, pp. 205–248. 1985.

Troutman, D.E., E.M. Godsy, D.F. Goerlitz and G.G. Ehrlich. *Phenolic Contamination in the Sand-and-Gravel Aquifer from a Surface Impoundment of Wood Treatment Wastewaters, Pensacola, Florida.* U.S. Geological Survey Water-Resources Investigations Report 84-4230. Prepared in cooperation with the Florida Department of Environmental Regulation. USGS, Denver, Colorado. 1984.

Varanasi, U. (Ed.). Metabolism of Polycyclic Aromatic Hydrocarbons in the Aquatic Environment. CRC Press. Boca Raton, Florida, 1989.

Vaughn, J.M., E.F. Landry, L.J. Baranosky, C.A. Beckwith, M.C. Dahl and N.C. Delihas. "Survey of Human Virus Occurrence in Wastewater Recharged Groundwater on Long Island." *Applied and Environmental Microbiology.* Volume 36, number 1, pp. 47–51. July, 1978.

Verschueren, K. *Handbook of Environmental Data on Organic Chemicals.* Second Edition, Van Nostrand Reinhold. New York, 1989.

Verschueren, K. *Handbook of Environmental Data on Organic Chemicals.* Second Edition, Van Nostrand Reinhold. New York, 1983.

Waller, Bradley, Barbara Howie and Carmen R. Causaras. *Effluent Migration from Septic Tank Systems in Two Different Lithologies, Broward County, Florida.* U.S. Geological Survey Water Resources Investigations Report 87-4075. Prepared in cooperation with Broward County. USGS, Denver, Colorado. 1987.

Wang, T.S., D.E. Spyridakis, B.W. Mar and R.R. Horner. *Transport, Deposition and Control of Heavy Metals in Highway Runoff.* WA-RD-39.10, U.S. Dept. of Transportation and Washington State Department of Transportation, Seattle, January 1980.

Wanielista, M.P., Y.A. Yousef and H.H. Harper. "Hydrology/Hydraulics of Swales," *Open Channel Hydraulics Workshop*, by the University of Central Florida and American Society of Civil Engineers, Florida Section, Orlando, Sept. 1983.

Wanielista, Martin P., Julius Charba, John Dietz, Richard S. Lott and Bryon Russell. *Evaluation of the Stormwater Treatment Facilities at the Lake Angel Detention Pond, Orange County, Florida.* State of Florida Department of Transportation Environmental Research Paper FL-ER-49-91. National Technical Information Service, Springfield, VA. 1991.

Wellings, F.M. "Perspective on Risk of Waterborne Enteric Virus Infection." *Chemical and Biological Characterization of Sludges, Sediments, Dredge Spoils, and Drilling Muds.* ASTM STP 976. American Society for Testing and Materials, Philadelphia, PA, pp. 257–264. 1988.

White, Everett M. and James M. Dornbush. "Soil Change Caused by Municipal Waste Water Applications in Eastern South Dakota." *Water Resources Bulletin.* Volume 24, number 2, pp. 269–273. April 1988.

Wilber, W.G. and J.V. Hunter. *The Influence of Urbanization on the Transport of Heavy Metals in New Jersey Streams.* Water Resources Research Institute, Rutgers University, New Brunswick, New Jersey, February 1980.

Wilde, F.D. *Geochemistry and Factors Affecting Ground-Water Quality at Three Storm-Water-Management Sites in Maryland: Report of Investigations No. 59.* Department of Natural Resources, Maryland Geological Survey, Baltimore, Maryland. (Prepared in Cooperation with the U.S. Department of the Interior Geological Survey, The Maryland Department of the Environment, and The Governor's Commission on Chesapeake Bay Initiatives). 1994.

Williams, L.H. "Effectiveness of Stormwater Detention." *Proceedings of the Conference on Stormwater Detention Facilities, Planning, Design, Operation, and Maintenance*, Henniker, New Hampshire, Edited by W. DeGroot, published by the American Society of Civil Engineers, New York, August 1982.

Wilson, L.G. "Sediment Removal from Flood Water." *Transactions American Society of Agricultural Engineers* 10(1):35–37, 1967.

Wilson, L.G., M.D. Osborn, K.L. Olson, S.M. Maida and L.T. Katz. "The Groundwater Recharge and Pollution Potential of Dry Wells in Pima County, Arizona." *Groundwater Monitoring Review*. Volume 10, pp. 114–121. Summer 1990.

Wolff, J., J. Ebeling and A. Muller. "Waste Water Irrigation Suited to the Environment as Shown by the Example of 'Abwasserverband Wolfsburg.'" *Hydrological Processes and Water Management in Urban Areas—Proceedings of the Conference*, International Hydrological Programme, UNESCO. pp. 599–607. Duisburg, Federal Republic of Germany. April 24–29, 1988.

Yousef, Y.A., M.P. Wanielista and H.H. Harper. "Design and Effectiveness of Urban Retention Basins." *Engineering Foundation Conference: Urban Runoff Quality—Impact and Quality Enhancement Technology*, Henniker, New Hampshire, edited by B. Urbonas and L.A. Roesner, published by the American Society of Civil Engineers, New York, June 1986.

Appendix A

ANNOTATED
BIBLIOGRAPHY OF
GROUNDWATER
CONTAMINATION

A bibliography of all of the literature sources used in the preparation of Chapter 3 of this book was assembled and is included in this appendix. This bibliography contains the abstracts of the various sources, as published in the original documents. A matrix of information (Table A-1) was also prepared. This matrix includes the author and the date of the reference in the same format that is used in the report. The matrix also notes the source water considered in the reference (stormwater, sanitary wastewater, industrial wastewater, or agricultural runoff), the location of the research (where applicable), and the contaminants discussed. The contaminants are organized in the same manner as in Chapter 3: nutrients, pesticides, organics, pathogens, metals, dissolved solids, suspended solids and dissolved oxygen. If a contaminant was addressed in the literature source, an "X" was placed in the appropriate column in Table A-1.

Alhajjar, B.J., G.V. Simsiman and G. Chesters. "Fate and Transport of Alachlor, Metolachlor and Atrazine in Large Columns." *Water Science and Technology: A Journal of the International Association of Water Pollution Research.* **V. 22, n. 6. pp. 87–94. 1990.**

^{14}C ring-labeled atrazine, alachlor, and metolachlor were surface-applied at 3.14 kg/ha in greenhouse lysimeters containing two soils in an ongoing experiment. Bromide (Br)—a conservative tracer—at 6.93 kg/ha as KBr and nitrate-nitrogen (NO_3-N) at 112

Table A.1. Groundwater Contamination References

Author (Date)	Source Water	Location of Work	Nutrients	Pesticides	Organics	Pathogens	Metals	Dissolved Solids	Suspended Solids	Dissolved Oxygen
Alhajjar, et al. (1990)	Agriculture			X						
Amer. Water Works Assoc. (1990)	Agriculture	Nebraska	X	X						
Amoros, et al. (1989)	Agriculture	Spain	X			X		X		
Armstrong and Llena (1992)	Stormwater	Wisconsin			X		X	X	X	
Aronson and Seaburn (1974)	Stormwater	Long Island, New York								
Asano (1987)	Sanitary	California	X			X	X	X	X	X
Bao-rui (1988)	Sanitary	China				X				
Barraclough (1966)	Industrial	Escam. & Santa Rosa Co., FL			X					
Boggess (1975)	Industrial	Lee County, FL	X					X		
Bouchard (1992)	Sanitary	Santa Monica, California					X			
Bouwer, et al. (1984)	Sanitary	Phoenix, Arizona			X					
Bouwer (1989)	Agriculture		X	X			X	X		
Bouwer (1987)	Agriculture		X	X				X		
Bouwer and Idelovitch (1987)	Sanitary	Arizona & Israel	X	X	X	X	X	X	X	
Brown (1982)	Sanitary	Tarpon Springs, Florida	X							
Butler (1987)	Stormwater	Austin, Texas	X	X	X	X		X		
Cavanaugh, et al. (1992)	Sanitary	North Carolina	X		X				X	
Chang, et al. (1988)	Sanitary	Riverside, California	X							X
Chase (1987)	Sanitary	Phoenix, Arizona								X

Reference	Type	Location
Cisic (1992)	Sanitary	
Close (1987)	Agriculture	
Clothier and Sauer (1988)	Agriculture	
Craun (1979)	Sanitary	
Crites (1985)	Sanitary	
Crook, et al. (1990)	Sanitary	California
Deason (1989)	Agriculture	CA & Western US
Deason (1987)	Agriculture	West & Midwest US
DeBoer (1983)	Sanitary	
Domagalski and Dubrovsky (1992)	Agriculture	San Joaquin Valley, CA
Ehrlich, et al. (1979a)	Industrial	Pensacola, Florida
Ehrlich, et al. (1979b)	Sanitary	Bay Park, New York
Elder, et al. (1985)	Sanitary	Tallahassee, Florida
Eren	Sanitary	Israel
Ferguson (1990)	Stormwater	Georgia
Ferguson, et al. (1990)	Agriculture	Nebraska
Gerba and Goyal (1985)	Sanitary	
Gerba and Haas (1988)	Sanitary	
German (1989)	Stormwater	Orlando, Florida
Goldschmid (1974)	Sanitary	Israel
Goolsby (1972)	Industrial	
Greene (1992)	Stormwater	Santa Monica, California
Hampson (1986)	Stormwater	Jacksonville, Florida
Harper (1988)	Stormwater	Florida
Hickey (1984)	Sanitary	St. Petersburg, Florida

Table A.1. Continued

Author (Date)	Source Water	Location of Work	Nutrients	Pesticides	Organics	Pathogens	Metals	Dissolved Solids	Suspended Solids	Dissolved Oxygen
Hickey and Vecchioli (1986)	Industrial	Florida			X					
Hickey and Wilson (1982)	Stormwater	Mulberry, Florida			X			X		
Higgins (1984)	Sanitary	New Jersey	X							
Horsley and Moser (1990)	Stormwater	Massachusetts	X	X			X	X	X	
Hull and Yurewicz (1979)	Stormwater	Live Oak, Florida	X	X	X	X	X			X
Ishizaki (1985)	Stormwater	Japan								
Jansons, et al. (1989a)	Sanitary					X				X
Jansons, et al. (1989b)	Sanitary				X	X		X		
Johnson (1987)	Sanitary	Tucson, Arizona								
Karkal (1992)	Sanitary	Orange Grove, California				X			X	X
Katapodes and Tang (1990)	Agriculture									
Kaufman (1973)	Sanitary	Florida	X		X			X		X
Knisel and Leonard (1989)	Agriculture	Georgia		X						
Krawchuk and Webster (1987)	Agriculture	Manitoba, Canada		X						
Ku and Simmons (1986)	Stormwater	Long Island, New York	X	X	X	X	X	X	X	X
Ku, et al. (1992)	Stormwater	Nassau County, New York	X		X	X	X	X	X	
Ku, et al. (1975)	Sanitary	Bay Park, New York	X		X			X		X
Lauer (1988a)	Agriculture	Patterson & Prosser, WA	X							

Reference	Type	Location
Lauer (1988b)	Agriculture	Patterson & Prosser, WA
Lee (1990)	Agriculture	
Lloyd, et al. (1988)	Stormwater	United Kingdom
Loh, et al. (1988)	Sanitary	Hawaii
Malik, et al. (1992)	Stormwater	Central Coast, California
Mancini and Plummer (1992)	Stormwater	
Markwood (1979)	Sanitary	
Marton and Mohler (1988)	Stormwater	Czechoslovakia
Marzouk, et al. (1979)	Sanitary	Israel
Merkel, et al. (1988)	Stormwater	Germany
Mossbarger and Yost (1990)	Agriculture	Corn Belt & Lake States, US
Nellor, et al. (1985)	Sanitary	Los Angeles County, CA
Nightingale, et al. (1983)	Agriculture	Fresno, California
Nightingale (1987a)	Stormwater	Fresno, California
Nightingale (1987b)	Stormwater	Fresno, California
Nightingale and Bianchi (1977a)	Agriculture	Fresno, California
Nightingale and Bianchi (1977b)	Agriculture	Fresno, California
Norberg-King, et al. (1991)	Agriculture	Colusa Basin Drain, CA
Pahren (1985)	Sanitary	
Peterson (1988)	Agriculture	Wyoming
Petrovic (1990)	Stormwater	
Phelps (1987)	Sanitary	Gainesville, Florida
Pierce and Wong (1988)	Agriculture	Canada

Table A.1. Continued

Author (Date)	Source Water	Location of Work	Nutrients	Pesticides	Organics	Pathogens	Metals	Dissolved Solids	Suspended Solids	Dissolved Oxygen
Pitt (1974)	Sanitary	Dade County, Florida	X		X	X	X	X	X	
Pitt, et al. (1975)	Sanitary	Dade County, Florida	X			X	X	X		X
Power and Schepers (1989)	Agriculture		X					X		
Pruitt, et al. (1985)	Sanitary	Florida		X	X					
Ragone (1977)	Sanitary	Bay Park, New York					X	X		X
Ragone, et al. (1975)	Sanitary	Bay Park, New York				X	X	X		X
Ragone and Vecchioli (1975)	Sanitary	Bay Park, New York	X				X	X		X
Ramsey, et al. (1987)	Sanitary	Lubbock, Texas	X			X		X		
Razack, et al. (1988)	Stormwater	France	X					X		
Rea and Istok (1987)	Agriculture									
Reichenbaugh (1977)	Agriculture	Lakeland, Florida	X			X				
Reichenbaugh, et al. (1979)	Sanitary	St. Petersburg, Florida	X			X		X		X
Rein, et al. (1992)	Sanitary		X			X				
Rice, et al. (1991)	Agriculture			X					X	
Ritter, et al. (1989)	Agriculture	Northeast & East US	X	X						
Ritter, et al. (1991)	Agriculture	Delaware	X							
Robinson and Snyder (1991)	Stormwater	South Carolina	X	X						
Rosenshein and Hickey (1977)	Sanitary	Pinellas Peninsula, Florida	X			X	X	X		
Sabatini and Austin (1988)	Agriculture			X						

Reference	Type	Location
Sabol, et al. (1987)	Agriculture	Arizona & New Mexico
Salo, et al. (1986)	Stormwater	California
Schiffer (1989)	Stormwater	Florida
Schmidt and Sherman (1987)	Agriculture	California
Schneider, et al. (1987)	Sanitary	Long Island, New York
Seaburn and Aronson (1974)	Stormwater	Long Island, New York
Shirmohammadi and Knisel (1989)	Agriculture	Southeast US
Smith and Myott (1975)	Sanitary	Long Island, New York
Spalding and Kitchen (1988)	Agriculture	Nebraska
Squires, et al. (1989)	Agriculture	California
Squires and Johnston (1989)	Agriculture	
Steenhuis, et al. (1988)	Agriculture	
Strutynski, et al. (1992)	Stormwater	
Tim and Mostaghimi (1991)	Sanitary	
Townley, et al. (1992)	Sanitary	Cambria, California
Treweek (1985)	Sanitary	Pensacola, Florida
Troutman, et al. (1984)	Industrial	Pensacola, Florida
US EPA (1992)	General	
Varuntanya and Shafer (1992)	Sanitary	Long Island, New York
Vaughn, et al. (1978)	Sanitary	
Vecchioli, et al. (1984)	Industrial	Pensacola, Florida

Table A.1. Continued

Author (Date)	Source Water	Location of Work	Nutrients	Pesticides	Organics	Pathogens	Metals	Dissolved Solids	Suspended Solids	Dissolved Oxygen
Verdin, et al. (1987)	Sanitary	Nevada	X							
Waller, et al. (1987)	Sanitary	Broward County, Florida						X	X	X
Wanielista, et al. (1991)	Stormwater	Orange County, Florida			X	X	X			
Wellings (1988)	Sanitary					X				
White and Dornbush (1988)	Sanitary	South Dakota	X					X		
Wilde (1994)	Stormwater	Maryland	X		X		X	X		
Wilson, et al. (1990)	Stormwater	Pima County, Arizona		X	X		X			
Wolff, et al. (1988)	Sanitary	Saxony, Germany	X				X	X		
Yurewicz and Rosenau (1986)	Sanitary	Tallahassee, Florida	X	X	X		X	X	X	X

kg/ha as KNO_3 were mixed with each herbicide and surface-applied. Growth of Red top (*Agrostis alba*) was established in each column (105 cm long and 29.4 cm i.d.). The experiment consisted of 12 columns (2 soils × 3 herbicides × 2 replicates) each fitted with four sampling ports for leachates, a volatilization chamber, and an aeration and irrigation system. Volatile materials are being trapped directly in solvents. One column replicate was dismantled for soil and plant analyses. Columns of Plainfield sand and Plano silt loam treated with alachlor and metolachlor were sampled after 23 and 28 weeks, respectively; the atrazine columns were sampled after 35 weeks. Herbicide residues are determined by liquid scintillation counting, extracted and separated by thin-layer chromatography using autoradiographic detection. Volatilization was ≤ 0.01% of the amount of herbicide applied. The order of herbicide mobility was alachlor > metolachlor >> atrazine. As many as 8 to 12 alachlor metabolites and 2 to 6 metolachlor metabolites were separated in leachates.

American Water Works Association. "Fertilizer Contaminates Nebraska Groundwater." *AWWA Mainstream.* **V. 34, n. 4. pp. 6. 1990.**

Nitrates and nitrogen from commercial fertilizers, manure, and other sources are increasingly contaminating large areas of groundwater in Nebraska. Nitrate was by far Nebraska's most frequently encountered contaminant. Most of the contamination stems from nonpoint sources such as rainwater runoff and erosion, not from spills or other accidents.

Amoros, I., J.L. Alonso and I. Peris. "Study of Microbial Quality and Toxicity of Effluents from Two Treatment Plants Used for Irrigation." *Water Science and Technology: A Journal of the International Association on Water Pollution Research.* **V. 21, n. 3. pp. 243–246. 1989.**

The use of treated wastewater for agriculture is currently practiced in many countries. The benefits of recycling wastewater are based upon its constituents, particularly nitrogen, phosphorus, and potassium, which are of value as fertilizers for many crops. There are, however, potential health risks associated with wastewater reuse. A number of bacterial pathogens excreted in the faeces of humans and animals are found in wastewater. The most significant pathogens, such as enteropathogenic *Escherichia coli, Salmonella, Shigella, Yersinia, Vibrio, Campylobacter,* and *Leptospira,* can be transmitted via sewage-irrigated vegetables. Populations of bacterial pathogens in wastewater may be quite high, although variable, and may depend on the effectiveness of wastewater treatment. In addition, toxic substances, such as heavy metals and some organic compounds, may be found.

In the province of Alicante, there are some arid agricultural areas designated for grape cultivation but which have water resource problems. The use of treated wastewater from secondary treatment plants is one solution to this problem. Some reservoirs have been constructed to store wastewater destined for irrigation.

A study of the health risks associated with wastewater reuse (especially important since grapes are destined for human consumption and are normally eaten raw, and not peeled) was carried out. The presence in water and on fruit of total coliforms, fecal coliforms and fecal streptococci, and pathogens such as *Salmonella* and *Vibrio cholerae* (only in wastewater) was investigated. Total and fecal coliforms and fecal streptococci were considered as indicator bacteria for the evaluation of the fecal pollution. *Salmonella* species were investigated because they are one of the pathogenic organisms most frequently found in sewage and they have been isolated from effluent samples after secondary treatment and disinfection with chlorine dioxide and ozone. *V. cholerae* is also transmitted via contaminated water, and some studies have reported prolonged survival of this microorganism in sewage with increased resistance to chlorine. Microbial toxicity tests to detect the potential toxicity of the wastewaters were conducted.

Armstrong, D.E. and R. Llena. *Stormwater Infiltration: Potential for Pollutant Removal.* **Project Report to the U.S. Environmental Protection Agency. Wisconsin Department of Natural Resources, University of Wisconsin Water Chemistry Program. 1992.**

The potential mobility of pollutants during infiltration of urban stormwater through soil was investigated. Groups of 32 organic and seven inorganic chemicals were evaluated. The framework for evaluating mobility was that of a conservative (nonsorbed) chemical. The main parameter controlling relative leaching rate was the pollutant K_d for sorption by the soil. For organic chemicals, K_d values were calculated based on predicted sorption to soil organic and inorganic components. Values were estimated for each compound in two soils at three organic matter contents. Soil organic matter was the dominant component controlling K_d, even in low-organic-matter soils. Among compounds, K_d increased with increasing hydrophobicity (octanol-water partition coefficient). For inorganic pollutants, published K_d values for two "typical" soils were used.

A mobility index (M_I = velocity of chemical/velocity of water under saturated flow) was used to classify pollutant leaching mobility. M_I values were calculated for each organic compound in two soils at three levels of organic concentrations. For organic chemicals, mobility was controlled by a combination of pollutant and soil properties. Organic matter content was the main soil property controlling mobility. In a low-organic-matter sandy loam soil (organic carbon = 0.01% by weight), ten compounds were classified as mobile (M_I = 0.1 to 1.0), while one was in the very low mobility group ($M_I < 0.001$). When soil organic carbon was increased to 1%, only five compounds remained in the high mobility group, and nineteen were in the very low mobility group. Hydrophobicity was the main organic chemical property controlling mobility. Pollutants remaining in the mobile and intermediate mobility groups even when soil organic matter was relatively high (1%) were characterized by octanol-water partition coefficients below 10^3.

The potential mobility of the inorganic pollutant group (mostly trace metals) was generally lower than for the organic chemical group. A nonlinear sorption model resulted in a predicted increase in mobility with concentration. At low concentration (0.01 mg/L), calculated mobilities were in the low (M_I = 0.001 to 0.01) or very low ($M_I < 0.001$) ranges.

Estimated residence times for leaching through the soil to the water table were estimated for a 1.0-meter soil layer and a water infiltration rate of about 60 m/yr. Residence times were influenced by pollutant mobility and infiltration site characteristics. Under the site conditions assumed, predicted residence times per meter were less than one year for mobile pollutants and less than 10 years for most low-mobility pollutants. Predicted residence times would increase directly with increasing soil depth or decreasing water loading rate. Residence times were calculated for a set of six pollutants at a hypothetical infiltration site in the Milwaukee, Wisconsin area.

Aronson, D.A. and G.E. Seaburn. *Appraisal of Operating Efficiency of Recharge Basins on Long Island, New York, in 1969.* **Geological Survey Water-Supply Paper 2001-D. U.S. Government Printing Office, Washington, D.C. 1974.**

Recharge basins on Long Island are unlined pits of various shapes and sizes excavated in surficial deposits of mainly glacial origin. Of the 2,124 recharge basins on Long Island in 1969, approximately 9% (194) contain water 5 or more days after a 1-inch rainfall. Basins on Long Island contain water because (1) they intersect the regional water table or a perched water table, (2) they are excavated in material of low hydraulic conductivity, (3) layers of sediment and debris of low hydraulic conductivity accumulate on the basin floor, or (4) a combination of these factors exists.

Data obtained as part of this study show that (1) 22 basins contain water because they intersect the regional water table, (2) a larger percentage of the basins excavated in the Harbor Hill and the Ronkonkoma morainal deposits contain water than basins excavated in the outwash deposits, (3) a larger percentage of basins that drain industrial and commercial areas contain water than basins that drain highways and residential areas, (4) storm runoff from commercial and industrial areas and highways generally contains high concentrations of asphalt, grease, oil, tar, and rubber particles, whereas runoff from residential areas mainly contains leaves, grass cuttings, and other plant material, and (5) differences in composition of the soils within the drainage areas of the basins on Long Island apparently are not major factors in causing water retention.

Water-containing basins dispose of an undetermined amount of storm runoff, primarily by the slow infiltration of water through the bottoms and the sides of the basins. The low average specific conductance of water in most such basins suggests that evaporation does not significantly concentrate the chemical constituents and, therefore, that evaporation is not a major mechanism of water disposal from these basins.

Asano, T. "Irrigation With Reclaimed Wastewater—Recent Trends." *Irrigation Systems for the 21st Century, Portland, Oregon, 1987.* **(American Society of Civil Engineers, New York, New York, pp. 735–742). 1987.**

Land application of municipal wastewater is a well established practice in California. Since approximately 33 million ac-ft (4×10^{10} m³) of water—or about 85% of the total water used in the state—are applied annually to approximately 9 million ac (36,400 km²) of irrigated cropland (California State Water Resources Control Board, 1981), irrigation with reclaimed wastewater has become a logical and important component of total water resources planning and development.

Much of the reclaimed municipal wastewater (57%) in California is used for irrigation of fodder, fiber, and seed crops (a use not requiring a high degree of wastewater treatment), and about 7% is used for irrigation of orchard, vine, and other food crops. An important use in recent years (about 14%) is irrigation of golf courses, other turfgrass, and landscape areas. According to a survey (Crook, 1985) conducted by the California Department of Health Services, approximately 271×10^6 m^3 of wastewater is reclaimed by 240 municipal wastewater treatment plants that supply reclaimed water to more than 380 use areas. Data from the 1984 survey on the types of reuse and numbers of use areas are listed.

Bao-rui, Y. "Investigation into Mechanisms of Microbial Effects on Iron and Manganese Transformations in Artificially Recharged Groundwater." *Water Science and Technology: A Journal of the International Association on Water Pollution Research*. V. 20, n. 3. pp. 47–53. 1988.

After artificial recharging of groundwater some problems occurred, such as changes in groundwater quality, the silting up of recharge (injection) wells, etc. Therefore, the mechanisms of microbial effects on groundwater quality after artificial recharging were studied in Shanghai and the district of Changzhou. These problems were approached on the basis of the amounts of biochemical reaction products generated by the metabolism of iron bacteria, sulfate-reducing bacteria, *Thiobacillus thioparus*, and *Thiobacillus denitrificans*. The experiments showed that in the transformations occurring and the siltation of recharge wells, microorganisms play an important role, due to the various chemical and biochemical activities. A water-rock-microorganisms system is proposed, and some methods for the prevention and treatment of these effects are given.

Barraclough, J.T. "Waste Injection into a Deep Limestone in Northwestern Florida." *Ground Water*. V. 3, n. 1. pp. 22–24. 1966.

During a three-month trial period, 70 million gallons of industrial wastes were successfully injected at moderate pressures into a deep limestone in the westernmost part of Florida. The movement of these wastes is expected to be predominantly southward toward the natural discharge area which is presumed to be far out in the Gulf of Mexico. The limestone lies between two thick beds of clay (aquicludes) and contains 13,000 ppm salty water. A series of aquifers and aquicludes appear capable of preventing contamination of the overlying freshwater aquifers.

Boggess, D.H. *Effects of a Landfill on Ground Water Quality*. Geological Survey Open File Report 75-594. U.S. Geological Survey, Tallahassee, Florida. 1975.

Chemical analyses of water from 11 wells were adequate to show definite effects on ground-water quality in the vicinity of a landfill site in Lee County, Florida, operated by the city of Fort Myers. These effects were observed as far as 1,200 ft (320 m) downgradient. When compared to the average concentrations of chemical constituents determined in comparable wells used as controls, water from the well with the greatest effect had sulfate 72 times greater, potassium 43 times greater, ammonia nitrogen 20 times greater, sodium and chloride 12 times greater, and most other chemical constituents 2 to 8 times greater.

The leachate was transported downgradient in the water table aquifer in the general direction of groundwater movement. A subsurface clay barrier and paths of higher permeability apparently caused some local variation in the direction of leachate movement. The leachate probably is transported more rapidly during high stages of the water table due to higher hydraulic gradients and increased transmissivity.

Because of the high water table, which is at or near the land surface during periods of maximum recharge from rainfall, extensive modification of the waste disposal procedures would be required to reduce or eliminate the effect of the leachate.

Bouchard, A.B. "Virus Inactivation Studies for a California Wastewater Reclamation Plant." *Water Environment Federation 65th Annual Conference & Exposition, New Orleans, Louisiana, 1992.* **(Water Environment Federation, Alexandria, Virginia). 1992.**

Engineering-Science, Inc. (ES) was retained by Thetford Systems, Inc. to develop, implement and supervise a virus inactivation study program for the Thetford Cycle-Let Wastewater Reclamation System. Thetford has installed a Cycle-Let system at the Water Garden Development Project in Santa Monica, California, to treat the collected municipal sewage from the project's office buildings and facilities. The Water Garden Cycle-Let system will produce reclaimed water for use as landscape irrigation for the development's property, and as makeup water for an onsite decorative lake. Phase II of the Water Garden Project proposes to use the reclaimed water as a source of toilet flush water in the office buildings. The State of California has established treatment process criteria required for certain types of reclaimed water use. Specifically, the California Administrative Code Title 22 identifies the processes required to treat municipal wastewater prior to reuse. Title 22 stipulates the following treatment processes to use reclaimed water as landscape irrigation—oxidation, clarification, coagulation, filtration, and disinfection using chlorination. The Thetford Cycle-Let extended aeration activated sludge with biological nitrification, followed by ultramembrane filtration, granular activated carbon adsorption, and ultraviolet light disinfection. As such, the California Department of Health Services (DOHS) has the requirements of Title 22. These requirements stipulate that the median number of coliform organisms in the plant effluent does not exceed 2.2 MPN per 100 mL, and that the effluent must be essentially pathogen free. ES was retained by Thetford to develop a DOHS-approved testing protocol, to implement the test program, to supervise the test program, and to prepare an engineering report to be submitted to the DOHS that demonstrates the adequacy of the extensive seeded virus inactivation study, followed by an in situ virus study, and the necessary bacteriological analyses. Since the Water Garden Project is still under construction and no wastewater is being currently generated, the study was conducted at a similar facility that serves an office park in Princeton, New Jersey. The effluent from the Cycle-Let system at this plant is used for toilet flush water in the office complex. By conducting the study at this plant, ES was able to test the Cycle-Let system's adequacy of meeting Title 22 and the possibility of using reclaimed water as toilet flush water for the Water Garden Project.

Bouwer, E.J., P.L. McCarty, H. Bouwer and R.C. Rice. "Organic Contaminant Behavior During Rapid Infiltration of Secondary Wastewater at the Phoenix 23rd Avenue Project." *Water Resource.* **V. 18, n. 4. pp. 463–472. 1984.**

Movement of trace organic pollutants during rapid infiltration of secondary wastewater for groundwater recharge was studied at the 23rd Avenue Rapid Infiltration Project in Phoenix, Arizona. Samples of the wastewater applied to the spreading basins and of renovated water taken from monitoring wells were characterized for priority pollutants and other specific organic compounds using gas chromatography/mass spectrometry. The concentrations of organic constituents were affected by volatilization, biodegradation, and sorption processes. Nonhalogenated aliphatics and aromatic hydrocarbons exhibited concentration decreases of 50–99% during soil percolation. Halogenated organic compounds were generally removed to a lesser extent. Concentrations of trichloroethylene, tetrachloroethylene, and pentachloroanisole appeared to be significantly higher in the renovated water than in the basin water; reasons for this behavior remain unclear. Many organic contaminants were detected in the groundwater, indicating that such systems should be designed to localize contamination of the aquifer. Chlorination of the wastewater had no significant effect on concentrations and types of trace organic compounds.

Bouwer, H. "Agricultural Contamination: Problems and Solutions." *Water Environment and Technology.* **V. 1, n. 2. pp. 292–297. 1989.**

Salts from irrigation water concentrate in the deep percolation water and can pollute groundwater, especially in dry climates where there is little natural dilution. Under certain geologic conditions, selenium and other trace elements may leach from the root and vadose zones into the groundwater. Salt contamination and trace element contamination are a direct result of agricultural activities, but neither is caused by anthropogenic chemicals, such as nitrate and pesticides, which can cause severe groundwater contamination.

Not long ago, it was thought that pesticides would not migrate to groundwater, except perhaps under situations of very coarse or cracked soils, shallow groundwater tables, high pesticide application rates, or accidental spills. In the mid to late 1970s, more and more cases of pesticide contamination were reported as well testing increased and more sensitive analytical equipment was used. The problem is now widespread. Many pesticides have been detected in groundwater, and thousands of wells have been closed.

Nitrogen-containing fertilizers also threaten groundwater quality because they produce nitrate in the soil. Nitrate moves readily with deep percolation through the vadose zone to groundwater. Thus, nitrogen not used by crops or denitrified in the soil and volatilized eventually appears as nitrate in groundwater. Nitrate concentrations in vadose zone water beneath agricultural fields typically range from 5 to 100 mg/L and are often 20 to 40 mg/L—10 to 30 mg/L greater than the maximum drinking water limit.

Growing concern over groundwater contamination caused by fertilizer and pesticide use has triggered an increase in legislative and regulatory activity. However, more research, including that on health effects and acceptable risks, is needed to establish sound regulations. Contaminant migration must also be better understood so that pesticide transport can be more accurately predicted. Preventing contamination is more effective than cleaning polluted aquifers, and for this purpose best management practices (BMPs) must be developed. Realistic regulatory policies and management practices that will protect public health while ensuring viable and sustainable agriculture must be implemented.

Bouwer, H. "Effect of Irrigated Agriculture on Groundwater." *Journal of Irrigation and Drainage Engineering*. V. 113, n. 1. pp. 4–15. 1987.

The time it takes for deep percolation water from irrigated fields to reach underlying groundwater increases with decreasing particle size of the vadose zone material and increasing depth to groundwater. For average deep percolation rates, decades may be required before the water joins the groundwater. Due to nonuniform irrigation applications and preferential flow, some deep percolation water will reach the groundwater much faster. Dissolved salts, nitrate, and pesticides are the chemicals in deep percolation water of main concern in groundwater pollution. Movement of pesticides may be retarded in the vadose zone, but biodegradation may also be slowed due to reduced organic carbon content and microbial activity at greater depths. Because of the large area of irrigated land in the world and the real potential for groundwater pollution, more research is necessary on downward movement of water and chemicals in the vadose zone.

Bouwer, H. and E. Idelovitch. "Quality Requirements for Irrigation with Sewage Water," *Journal of Irrigation and Drainage Engineering*. V. 113, n. 4. pp. 516–535. 1987.

Irrigation is an excellent use for sewage effluent because it is mostly water with nutrients. For small flows, the effluent can be used on special, well supervised "sewage farms," where forage, fiber, or seed crops are grown that can be irrigated with standard primary or secondary effluent. Large-scale use of the effluent requires special treatment so that it meets the public health, agronomic, and aesthetic requirements for unrestricted use (no adverse effects on crops, soils, humans, and animals). Crops in the unrestricted-use category include those that are consumed raw or brought raw into the kitchen. Most state or government standards deal only with public health aspects, and prescribe the treatment processes or the quality parameter that the effluent must meet before it can be used to irrigate a certain category of crops. However, agronomic aspects related to crops and soils must also be taken into account. Quality parameters to be considered include bacteria, viruses, and other pathogens; total salt content and sodium adsorption ratio of the water (soil as well as crop effects); nitrogen; phosphorus; chloride and chlorine; bicarbonate; heavy metals, boron, and other trace elements; pH; and synthetic organics (including pesticides).

Brown, D.P. *Effects of Effluent Spray Irrigation on Ground Water at a Test Site Near Tarpon Springs, Florida*. Geological Survey Open-File Report 81-1197. U.S. Geological Survey, Tallahassee, Florida. 1982.

Secondary-treated effluent was applied to a 7.2-ac test site for about 1 year at an average rate of 0.06 million gallons per day and 3 years at 0.11 million gallons per day.

Chemical fertilizer was applied periodically to the test site and adjacent areas. Mounding of the water table occurred due to effluent irrigation, inducing radial flow from the test site.

Ground water in the surficial aquifer at the test site and adjacent areas showed substantial increases in most chemical and physical parameters (including chloride, specific conductance, pH, total nitrogen, and total carbon) above the range of values observed in nearby areas that were irrigated with water from the Floridan aquifer and periodically fertilized.

In the surficial aquifer, about 200 ft downgradient from the test site, physical, geochemical, and biochemical processes effectively reduced total nitrogen concentration 90% and total phosphorus concentration more than 95% from that of the applied effluent. In the effluent, total nitrogen averaged 26 mg/L and total phosphorus averaged 7.3 mg/L. Downgradient, total nitrogen averaged 2.4 mg/L and total phosphorus averaged 0.17 mg/L. Increases in total phosphorus concentration were observed where the pH of groundwater increased.

Microbiological data did not indicate fecal contamination in the surficial aquifer. Fecal coliform bacteria were generally less than 25 colonies per 100 mL at the test site and were not detected downgradient or near the test site. Fecal streptococcal bacteria were generally less than 100 colonies per 100 mL at the test site and on three occasions were detected adjacent to the test site. In the Floridan aquifer, total and fecal coliform bacteria were detected in 50% of the samples. Total coliform bacteria were generally less than 100 colonies per 100 mL and fecal coliform bacteria were generally less than 10 colonies per 100 mL.

Butler, K.S. "Urban Growth Management and Groundwater Protection: Austin, Texas." *Planning for Groundwater Protection.* **Academic Press, New York, New York. pp. 261–287. 1987.**

Austin, Texas has recognized that the greatest opportunity to prevent declining groundwater quality is through proper location, design, construction, and maintenance of new urban development and its associated drainage systems. Increasing development in the environmentally sensitive watersheds south and west of Austin resulted in the promulgation of a series of innovative watershed development ordinances in the early 1980s (Butler, 1983). These regulations are designed to protect the water quality of the Edwards Aquifer, a unique karstic limestone system. This process of planning for the protection of groundwater is the subject of this report.

Specifically, this report concerns the effects that the ordinances for several watersheds and the associated aquifer will have on suburban land development and water quality protection for the Edwards Aquifer. These ordinances are salient examples of a new, largely untested venture into urban runoff pollution control using a combination of engineering and land use management techniques. The immediate motivation for adopting these four techniques is protecting groundwater and spring discharges from degradation and contamination with nontoxic chemicals. Austin also recognized that toxic contaminants may become a threat to groundwater. The combination of engineering and land use management techniques as adopted in Austin's ordinances should protect the groundwater from organic and inorganic microcontaminants as well as more common pollutants associated with urban storm runoff.

This report addresses three key questions: Why is the Edwards Aquifer area so deserving of special protection in the face of urban expansion? How do these development standards affect the planning of new subdivisions and site developments? And how do groundwater quality protection standards operate in the broader context of growth management in this rapidly urbanizing region of central Texas?

Cavanaugh, J.E., H.S. Weinberg, A. Gold, R. Sangalah, D. Marbury, W.H. Glaze, T.W. Collette, S.D. Richardson and A.D. Thurston Jr. "Ozonation Byproducts: Identification of Bromohydrins from the Ozonation of Natural Water with Enhanced Bromide Levels." *Environmental Science & Technology.* **V. 26, n. 8. pp. 1658–1662. 1992.**

When ozone is used in the treatment of drinking water, it reacts with both inorganic and organic compounds to form byproducts. If bromide is present, it may be oxidized to hypobromous acid, which may then react with natural organic matter (NOM) to form brominated organic compounds. The formation of bromoform has been well documented, and more recently, other byproducts—such as bromoacetic acids, bromopicrin, cyanogen bromide, bromoacetones, and bromate—have been identified. The purpose of this communication is to report the identification of bromohydrins, a new group of labile brominated organic byproducts from the ozonation of a natural water in the presence of enhanced levels of bromide.

Chang, A.C., A.L. Page, P.F. Pratt and J.E. Warneke. "Leaching of Nitrate from Freely Drained-Irrigated Fields Treated with Municipal Sludges." *Planning Now for Irrigation and Drainage in the 21st Century, Lincoln, Nebraska, 1988.* **(American Society of Civil Engineers, New York, New York, pp. 455–467). 1988.**

A municipal sludge land application experiment was initiated in 1975 near Riverside, California. From 1975 to 1983 sludges from the Los Angeles Metropolitan Area were applied on experimental plots where crops were grown annually. Results indicated that (1) crop yields were not affected by the sludge application and the total biomass harvested was a function of total nitrogen input, (2) the extent and the rate of nitrogen mineralization of liquid sludges were consistently higher than composted sludges, and the sludge application rate and the soil type did not affect mineralization of N, and (3) nitrogen application exceeding nitrogen requirements for crop growth always results in leaching and accumulation of nitrate in the soil profile.

Chase, W.L.J. "Reclaiming Wastewater in Phoenix, Arizona." *Irrigation Systems for the 21st Century, Portland, Oregon, 1987.* **(American Society of Civil Engineers, New York, New York, pp. 336–343). 1987.**

This paper examines the role of wastewater effluent reuse in the future water resource management of the City of Phoenix. The paper seeks to explain why a proposal to renovate effluent through rapid infiltration land treatment and to recycle that renovated water for agricultural irrigation has still not been implemented 20 years after the first serious studies of this project were initiated by the Salt River Project and the U.S. Department of Agriculture's Water Conservation Laboratory in Phoenix. While answering the technical questions concerning the safety of using renovated wastewater effluent for unrestricted agricultural irrigation accounted for a little of over half of that time, the primary problems since 1982 have been economic, legal, and institutional.

Cisic, S., D. Marske, B. Sheikh, B. Smith and F. Grant. "City of Los Angeles Effluent—Today's Waste, Tomorrow's Resource." *Water Environment Federation 65th Annual Conference & Exposition, New Orleans, Louisiana, 1992.* **(Water Environment Federation, Alexandria, Virginia). 1992.**

The City of Los Angeles receives water from three sources. In an average year, 70% of the water used to come from the Eastern Sierra Nevada via the Los Angeles Aqueduct; wells in the San Fernando Valley and other local groundwater basins supplied 16%; and purchases from the Metropolitan Water District of Southern California provided the remaining 14%. Faced with the fifth consecutive year of drought and supply cut from all three present sources, the City is planning to tap into a fourth—its wastewater.

Close, M.E. "Effects of Irrigation on Water Quality of a Shallow Unconfined Aquifer." *Water Resources Bulletin.* **V. 23, n. 5. pp. 793–802. 1987.**

The groundwater quality of a shallow unconfined aquifer was monitored before and after implementation of a border strip irrigation scheme, by taking monthly samples from an array of 13 shallow wells. Two 30-m-deep wells were sampled to obtain vertical concentration profiles. Marked vertical, temporal, and spatial variabilities were recorded. The monthly data were analyzed for step and linear trends using nonparametric tests that were adjusted for the effects of serial correlation. Average nitrate concentrations increased in the preirrigation period and decreased after irrigation began. This was attributed to wetter years in 1978–1979 than 1976–1977 which increased leaching, and to disturbance of the topsoil during land contouring before irrigation, followed by excessive drainage after irrigation. Few significant trends were recorded for other determinants, possibly because of shorter data records.

Nitrate, sulfate, and potassium concentrations decreased with depth, whereas sodium, calcium, bicarbonate, and chloride concentrations increased. These trends allowed an estimation to be made of the depth of groundwater affected by percolating drainage. This depth increased during the irrigation season and after periods of winter recharge. Furthermore, an overall increase in the depth of drainage-affected groundwater occurred with time, which paralleled the development of the irrigation scheme.

Clothier, B.E. and T.J. Sauer. "Nitrogen Transport during Drip Fertigation with Urea." *Soil Science Society of America Journal.* **V. 52, n. 2. pp. 345–349. 1988.**

Urea added to drip irrigation water will be rapidly hydrolyzed in the soil to ammonium and then oxidized to nitrate. An approximate theory is presented for the unsteady, three-dimensional transport of water and nitrogen through unsaturated soil around a dripper discharging a urea solution. The results were compared with measurements from laboratory experiments with repacked silt loam. Water and solute movement in the course of the irrigation cycle and during the subsequent redistribution are considered. The theory successfully located the penetration of both the inert nitrate and reactive ammonium derived from the applied urea. It was possible to predict the direction and approximate magnitude of pH changes proximal to the emitter.

Craun, G.F. "Waterborne Disease—A Status Report Emphasizing Outbreaks in Ground-Water Systems." *Ground Water.* **V. 17, n. 2. pp. 183–191. 1979.**

A total of 192 outbreaks of waterborne disease affecting 36,757 persons were reported in the United States during the period 1971–1977. More outbreaks occurred in nonmunicipal water systems (70%) than municipal water systems; however, more illness (67%) resulted from outbreaks in municipal systems. Almost half of the outbreaks (49%) and illness (42%) were caused by either the use of untreated or inadequately treated groundwater. An unusually large number of waterborne outbreaks affected travelers, campers, visitors to recreational areas, and restaurant patrons during the months of May through August and involved nonmunicipal water systems which primarily depend on groundwater sources. The major causes of outbreaks in municipal systems were contamination of the distribution systems and treatment deficiencies, which accounted for 68% of the outbreaks and 75% of the illness that occurred in municipal systems. Use of untreated groundwater was responsible for only 10% of the municipal system outbreaks and 1% of the illness. The major cause of the outbreaks in nonmunicipal systems was use of untreated groundwater which accounted for 44% of the outbreaks and 44% of the illness in these systems. Treatment deficiencies, primarily inadequate and interrupted chlorination of groundwater sources, were responsible for 34% of the outbreaks and 50% of the illness in nonmunicipal water systems.

Crites, R.W. "Micropollutant Removal in Rapid Infiltration." *Artificial Recharge of Groundwater.* **Butterworth Publishers, Boston. pp. 579–608. 1985.**

In a rapid-infiltration land treatment system, wastewater is treated as it percolates through the soil. The wastewater is applied to moderately and highly permeable soils (such as sands) by surface spreading in level basins or by sprinkling. Treatment is accomplished by biological, physical, and chemical means within the soil.

The need for definitive information on the extent of soil treatment during rapid infiltration has been recognized. Rapid infiltration is effective in removing many wastewater constituents—such as suspended solids, BOD_5, ammonium-nitrogen, phosphorus, bacteria, and virus—and is less effective in removing other constituents, such as nitrate-nitrogen, trace organics, and trace minerals.

The constituents addressed in this report include trace organics, inorganics (those covered by drinking water standards), and microorganisms. For these three classes of constituents, the health effects, removal mechanisms, and removals in existing rapid infiltration systems are discussed.

Crook, J., T. Asano and M. Nellor. "Groundwater Recharge with Reclaimed Water in California." *Water Environment & Technology.* **V. 2, n. 8. pp. 42–49. 1990.**

In California, increasing demands for water have given rise to surface water development and large-scale projects for water importation. Economic and environmental concerns associated with these projects have expanded interest in reclaiming municipal wastewater to supplement existing water supplies. Groundwater recharge represents a large potential use of reclaimed water in the state. For example, several projects have been identified in the Los Angeles area that could use up to 150×10^6 m³/ha (120,000 ac-ft/yr) of reclaimed water for groundwater recharge. Recharging groundwater with reclaimed wastewater has several purposes: to prevent saltwater intrusion into freshwater aquifers, to store the reclaimed water for future use, to control or prevent ground subsidence, and to augment nonpotable or potable groundwater aquifers. Recharge can be accomplished by surface spreading or direct injection.

With surface spreading, reclaimed water percolates from spreading basins through an unsaturated zone to the groundwater. Direct injection entails pumping reclaimed water directly into the groundwater, usually into a confined aquifer. In coastal areas, direct injection effectively creates barriers that prevent saltwater intrusion. In other areas, direct injection may be preferred where groundwater is deep or where the topography or existing land use makes surface spreading impractical or too expensive. While only two large-scale, planned operations for groundwater recharge are using reclaimed water in California, incidental or unplanned recharge is widespread.

The constraints of groundwater recharge with reclaimed water include water quality, the potential for health hazards, economic feasibility, physical limitations, legal restriction, and the availability of reclaimed water. Of these concerns, the health concerns are by far the most important, as they pervade all potential recharge projects. Health authorities emphasize that indirect potable reuse of reclaimed wastewater through groundwater recharge encompasses a much broader range of potential risks to the public's health than nonpotable uses of reclaimed water. Because the reclaimed water eventually becomes drinking water and is consumed, health effects associated with prolonged exposure to low levels of contaminants and acute health effects from pathogens or toxic substances must be considered. Particular attention must be given to organic and inorganic substances that may elicit adverse health responses in humans after many years of exposure.

Deason, J.P. "Irrigation-Induced Contamination: How Real a Problem?" *Journal of Irrigation and Drainage Engineering.* **V. 115, n. 1. pp. 9–20. 1989.**

The U.S. Department of the Interior has embarked on a series of reconnaissance level investigations throughout the western states to identify, evaluate, and respond to irrigation-induced water quality problems. A series of water, sediment, and biological samples are being analyzed for 17 inorganic constituents and a number of pesticides. Nineteen studies in 13 states have been undertaken. Seven have been completed to date. Results of the seven studies that have been completed are presented and compared to baselines, standards, criteria, and other guidelines helpful for assessing the potential of observed constituent concentrations in water, bottom sediment, and biota, to result in physiological harm to fish, wildlife, or humans. These initial results indicate that a new environmental problem of major proportions does not exit, but that some localized problems of significant magnitude do exist and should be addressed.

Deason, J.P. "Selenium: It's Not Just in California." *Irrigation Systems for the 21st Century, Portland, Oregon, 1987.* **(American Society of Civil Engineers, New York, New York, pp. 475–482). 1987.**

The U.S. Department of the Interior has embarked on a series of reconnaissance level investigations throughout the western states to identify and assess potential irrigation-induced water quality problems. A series of water, sediment and biological samples are being analyzed for selenium and 16 other trace elements, as well as a number of pesticides. Nine studies in seven states are underway currently and ten additional locations for study have been identified.

DeBoer, J.G. "Wastewater Reuse: A Resource or a Nuisance?" *Journal of the American Water Works Association.* **V. 75. pp. 348–356. 1983.**

As demand for good quality water increases, some areas in the United States find that traditional finite sources cannot meet all needs. Although many utilities draw potable supplies from water that has previously been used upstream, the direct reuse of treated water for potable supplies is limited by costs and by unknowns concerning health effects. Planned reuse options, however, are feasible water conservation techniques, especially for industrial and agricultural uses. Reclaimed wastewater can also be used to recharge groundwater, thereby augmenting potable supplies.

Domagalski, J.L. and N.M. Dubrovsky. "Pesticide Residues in Ground Water of the San Joaquin Valley, California." *Journal of Hydrology.* **V. 130, n. 1–4. pp. 299–338. 1992.**

A regional assessment of nonpoint source contamination of pesticide residues in groundwater was made of the San Joaquin Valley, an intensively farmed and irrigated structural trough in central California. About 10% of the total pesticide use in the USA is in the San Joaquin Valley. Pesticides detected include atrazine, bromacil, 2,4-DP, diazinon, dibromochloropropane, 1,2-dibromoethane, dicamba, 1,2-dichloropropane, diuron, prometon, prometryn, propazine, and simazine. All are soil applied except diazinon.

Pesticide leaching is dependent on use patterns, soil texture, total organic carbon in soil, pesticide half-life, and depth to water table. Leaching is enhanced by flood-irrigation methods, except where the pesticide is foliar applied, such as diazinon. Soils in the western San Joaquin Valley are fine grained and are derived primarily from marine shales of the Coast Ranges. Although shallow groundwater is present, the fewest number of pesticides were detected in this region. The fine-grained soil inhibits pesticide leaching because of either low vertical permeability or high surface area; both enhance adsorption onto solid phases. Soils in the eastern part of the valley are coarse grained with low total organic carbon and are derived from Sierra Nevada granites. Most pesticide leaching is in these alluvial soils, particularly in areas where depth to groundwater is less than 30 m. The areas currently most susceptible to pesticide leaching are eastern Fresno and Tulare Counties.

Tritium in water molecules is an indicator of aquifer recharge with water of recent origin. Pesticide residues transported as dissolved species were not detected in nontritiated water. Although pesticides were not detected in all samples containing high tritium, these samples are indicative of the presence of recharge water that interacted with agricultural soils.

Ehrlich, G.G., E.M. Godsy, C.A. Pascale, and J. Vecchioli. "Chemical Changes in an Industrial Waste Liquid During Post-Injection Movement in a Limestone Aquifer, Pensacola, Florida." *Ground Water*. V. 17, n. 6. pp. 562–573. 1979a.

An industrial waste liquid containing organonitrile compounds and nitrate ion has been injected into the lower limestone of Floridan aquifer near Pensacola, Florida since June 1975. Chemical analyses of water from monitor wells and backflow from the injection well indicate that organic carbon compounds are converted to CO_2 and nitrate is converted to N_2. These transformations are caused by bacteria immediately after injection, and are virtually completed within 100 m of the injection well. The zone near the injection well behaves like an anaerobic filter, with nitrate-respiring bacteria dominating the microbial flora in this zone.

Sodium thiocyanate contained in the waste is unaltered during passage through the injection zone and is used to detect the degree of mixing of injected waste liquid with native water at a monitor well 312 m (712 ft) from the injection well. The dispersivity of the injection zone was calculated to be 10 m (33 ft). Analyses of samples from the monitor well indicate 80% reduction in chemical oxygen demand and virtually complete loss of organonitriles and nitrate from the waste liquid during passage from the injection well to the monitor well. Bacteria densities were much lower at the monitor well than in backflow from the injection well.

Ehrlich, G.G., H.F.H. Ku, J. Vecchioli and T.A. Ehlke. *Microbiological Effects of Recharging the Magothy Aquifer, Bay Park, New York, with Tertiary-Treated Sewage*. Geological Survey Professional Paper 751-E. U.S. Government Printing Office, Washington, D.C. 1979b.

Injection of highly treated sewage (reclaimed water) into a sand aquifer on Long Island, NY, stimulated microbial growth near the well screen. Chlorination of the injectant to 2.5 mg/L suppressed microbial growth to the extent that it did not contribute significantly to head buildup during injection. In the absence of chlorine, microbial growth caused extensive well clogging in a zone immediately adjacent to the well screen.

During a resting period of several days between injection and well redevelopment, the inhibitory effect of chlorine dissipated, and microbial growth ensued. The clogging mat at the well/aquifer interface was loosened during this period, probably as a result of microbial activity.

Little microbial activity was noted in the aquifer beyond 20 ft from the well screen; this activity probably resulted from small amounts of biotransformable substances not completely filtered out of the injectant by the aquifer materials.

Movement of bacteria from the injection well into the aquifer was not extensive. In one test, in which injected water had substantial total coliform, fecal coliform, and fecal streptococcal densities, no fecal coliform or fecal streptococcal bacteria, and only nominal total coliform bacteria, were found in water from an observation well 20 ft from the point of injection.

Elder, J.F., J.D. Hunn and C.W. Calhoun. *Wastewater Application by Spray Irrigation on a Field Southeast of Tallahassee, Florida: Effects on Ground-Water Quality and Quantity.* **Geological Survey Water-Resources Investigations Report 85-4006. U.S. Geological Survey, Tallahassee, Florida. 1985.**

An 1,840-ac agriculture field southeast of Tallahassee, Florida, that has been used for land application of wastewater by spray irrigation is the site of a long-term groundwater monitoring study. The purpose of the study is to determine effects of wastewater application on water table elevations and groundwater quality. The study was conducted in cooperation with the City of Tallahassee. This report summarizes the findings for the period 1980–82.

Wastewater used for spray irrigation has high concentrations, relative to those in groundwater, of chloride, nitrogen, phosphorus, organic carbon, coliform bacteria, sodium, and potassium. At most locations, percolation through the soil has been quite effective in attenuation of these substances before they can impact the groundwater. However, increases in chloride and nitrate-nitrogen were evident in groundwater in some of the monitoring wells during the study, especially those wells which are within the sprayed areas. Chloride concentrations, for example, increased from approximately 3 mg/L to 15 to 20 mg/L in some wells and nitrate-nitrogen concentrations increased less than 0.5 mg/L to 4 mg/L or more.

Groundwater levels in the area of the spray field fluctuated over a range of several feet. These fluctuations were affected somewhat by spray irrigation, but the primary control on water levels was rainfall.

As of December 1982, constituents introduced to the system by spray irrigation of effluent had not exceeded drinking water standards in the groundwater. The system has not yet stabilized, however, and more changes in groundwater quality may be expected.

Eren, J. Mekoroth Water Co., Northern District, P.O.B. 755, Haifa, Israel. "Changes in Wastewater Quality during Long Term Storage."

The rainfall in Israel is in the winter months—November through April—and the irrigation period is during the summer months—May through October. Cotton, which is the main crop irrigated by reused wastewater, requires irrigation during three months (June through August) only.

Therefore, wastewater reuse projects require facilities to store the winter effluents for summer utilization.

It is either underground storage in sandy aquifers—as in the Dan Region reclamation project—or storage in deep wastewater reservoirs, which are used in all the other projects. The largest of these reservoirs is the Maale Kishon (Upper Kishon) Reservoir, which is part of the Kishon Project, which reuses wastewater from the Haifa Metropolitan Region.

The storage duration is from a few weeks up to several months, and during this period there is intense biological activity, which causes significant changes in water quality in the reservoirs. In a preliminary study in a small reservoir (1) prior to the construction of the Maale Kishon Reservoir it was observed that 6 weeks of storage considerably improved the water quality. The purpose of this investigation was to determine the changes that occur in a relatively large water body and how those affect the parameters that are important for agricultural irrigation.

Ferguson, B.K. "Role of the Long-Term Water Balance in Management of Stormwater Infiltration." *Journal of Environmental Management.* **V. 30, n. 3. pp. 221–233. 1990.**

Artificial infiltration of urban stormwater can potentially recharge groundwater and sustain stream base flows while improving stormwater quality and contributing to flood control. It involves capturing stormwater in basins where it is stored while infiltrating the surrounding soil. This paper suggests that management of these basins with design storm approaches needs to be supplemented by the long-term water balance incorporating continuous low-level background flows, in contrast to the design storm, which is an isolated, rare, and brief event. Background flows can accumulate in basins with no regular surface outlets, potentially reducing basin capacity and causing nuisances associated with standing water. A model is described for routing monthly average flows through infiltration basins. Using this model, 12 infiltration basins representing different construction methods and management objectives were designed for a hypothetical catchment in the Atlanta area. The effects of these basins in terms of cost, presence of standing water, capture of flood flows and average annual disposition of water were evaluated. The results show that background flows cannot be disregarded in infiltration management, since the performance of basins designed without considering background flows can be considerably hampered by their presence. The results also invite discussion of alternative basin geometries, materials, and hydroperiods as ways of meeting site-specific objectives for water resources and urban amenities.

Ferguson, R.B., D.E. Eisenhauer, T.L. Bockstadter, D.H. Krull and G. Buttermore. "Water And Nitrogen Management in Central Platte Valley of Nebraska." *Journal of Irrigation and Drainage Engineering.* **V. 116, n. 4. pp. 557–565. 1990.**

Contamination of groundwater by nitrogen leached from fertilizer on irrigated soils is related to the quantity of nitrate-N (NO_3^--N) present, the leaching potential based on soil texture and percentage depletion of available soil-water in the root zone, and the amount of water entering the soil profile. Research and demonstration projects in the central Platte valley of Nebraska have shown that NO_3^--N leaching is influenced by both irrigation and fertilizer-nitrogen (N) management in corn production. Scheduling irrigation according to available soil-water depletion can reduce deep percolation to a certain extent. Additional reduction in deep percolation can be achieved by improving efficiency of water application, particularly on furrow irrigated fields. Testing for NO_3^--N in irrigation water and soil can provide for substantial reductions in fertilizer nitrogen application, if residual levels in the soil are high, or if considerable NO_3^--N will be applied with irrigation water. Grain yields were not appreciably affected by the use of these management practices, while in most cases input costs for fertilizer nitrogen and irrigation water were reduced.

Gerba, C.P. and S.M. Goyal. "Pathogen Removal from Wastewater During Groundwater Recharge." *Artificial Recharge of Groundwater.* **Butterworth Publishers, Boston. pp. 283–317. 1985.**

Groundwater contamination by pathogenic microorganisms has not received as much attention as surface water pollution because it is generally assumed that groundwater has a good microbiologic quality and is free of pathogenic microorganisms. A number of well documented disease outbreaks have, however, been traced to contaminated groundwater. A total of 673 waterborne outbreaks affecting 150,268 persons occurred in the United States from 1946 to 1980. Of these, 295 (44%) involving 65,173 cases were attributed to contamination of groundwater.

Currently, 20% of the total water consumed in the United States is drawn from groundwater sources, and it is estimated that this usage will increase to 33% in the year 2000. According to Duboise et al. over 60 million people in the United States are served by public water supplies using groundwater, and about 54% of rural population and 2% of the urban population obtain their water from individual wells. Since groundwater is often used for human consumption without any treatment, it is imperative to understand the fate of pathogenic microorganisms during land application of wastewater.

The sources of fecal contamination may include septic tanks, leaky sewer lines, lagoons and leaching ponds, sanitary landfills for solid wastes, and sewage oxidation ponds. Additional sources of pathogens in groundwater may involve artificial recharge of groundwater aquifers with renovated wastewater including deep well injection, spray irrigation of crops and landscape, basin recharge, and land application of sewage effluent and sludges. Leakage of sewage into the groundwater from septic tanks, treatment lagoons and leaky sewers is estimated to be over a trillion gallons a year in the United States.

It should be realized that, as opposed to surface water pollution, contamination of groundwater is much more persistent and is difficult to eradicate. Because restoration of groundwater quality is difficult, time-consuming, and expensive, efforts should be made for the protection of groundwater quality rather than only for its restoration after degradation.

Gerba, C.P. and C.N. Haas. "Assessment of Risks Associated with Enteric Viruses in Contaminated Drinking Water." *Chemical and Biological Characterization of Sludges, Sediments, Dredge Spoils, and Drilling Muds.* **ASTM STP 976. American Society for Testing and Materials, Philadelphia, Pennsylvania. pp. 489–494. 1988.**

It is now well established that enteric viruses, such as hepatitis A, Norwalk, rotavirus, and so forth, can be transmitted by sewage-contaminated water and food. Standards for viruses in water have been suggested by the World Health Organization and several other organizations. Few attempts have been made to assess the risks associated with exposure to low numbers of enteric viruses in the environment.

To determine the risks that may be associated with exposure to human enteric viruses, the literature on minimum infectious dose, incidence of clinical illness, and mortality was reviewed. This information was then used to assess the probability of infection, illness, and mortality for individuals consuming drinking water containing various concentrations of enteric viruses. Risks were determined on a daily, annual, and lifetime basis. This analysis suggested that significant risks of illness ($>1:10^6$) may arise from exposure to low levels of the enteric virus.

German, E.R. *Quantity and Quality of Stormwater Runoff Recharged to the Floridan Aquifer System Through Two Drainage Wells in the Orlando, Florida, Area.* Geological Survey Water Supply Paper 2344. U.S. Government Printing Office, Washington, D.C. 1989.

Quantity and quality of inflow to two drainage wells in the Orlando, Florida, area were determined for the period April 1982 through March 1983. The wells, located at Lake Midget and at Park Lake, are used to control the lake levels during rainy periods. The lakes receive stormwater runoff from mixed residential-commercial areas of about 64 ac (Lake Midget) and 96 ac (Park Lake) and would frequently flood adjacent areas if the wells did not drain the excess stormwater. These lakes and wells are typical of stormwater drainage systems in the area.

Lake stages were monitored and used to estimate quantities of drainage well inflow. Estimated inflow for April 1982 through March 1983 was 62.4 ac-ft at Lake Midget and 84.0 ac-ft at Park Lake. Inflow to the drainage wells was sampled periodically. The quality of water prior to inflow to the drainage wells was estimated from samples of stormwater runoff to the lakes. The quality of formation water near the wells was estimated from samples pumped from the two drainage wells. A reconnaissance sampling of inflow at seven other drainage wells in the Orlando area was done, once at each well, to broaden the areal coverage of the investigation. The laboratory analyses included determinations of selected nutrients, bacteria, major constituents, trace elements, and numerous organic compounds, including many designated priority pollutants by the U.S. Environmental Protection Agency.

Comparison of quality of drainage well inflow with State criteria for drinking water supply indicated that color and bacteria were excessively high, and pH excessively low, in some samples. Constituents that exceeded the criteria were iron, in 10 to 21 inflow samples, manganese, in 1 sample, and lead, in 1 sample.

The priority pollutant dibenzo (a,h) anthracene was present in one of two samples pumped from the Lake Midget drainage well (concentration of 370 mg/L). The presence of this compound in that high a concentration is puzzling, because it was not detected in any samples of stormwater runoff or drainage well inflow.

Pesticides—especially diazinon, malathion, and 2,4-D—were the most frequently detected organic compounds in stormwater runoff, drainage well inflow, and Floridan aquifer system water samples. The priority pollutant bis (2-ethylhexyl) phthalate was detected in seven samples from five sites, probably because of widespread use of the compound in plastic products. Polynuclear aromatic compounds (fluoranthene, pyrene, anthracene, and chrysene) were found in stormwater runoff or inflow to drainage wells at Lake Midget and Park Lake and may be associated with runoff containing petroleum products.

Estimated annual loads to the Floridan aquifer system through drainage well inflow in the Orlando area, in pounds, are: dissolved solids, 32,000,000; total nitrogen, 100,000; total phosphorus, 13,000; total recoverable lead, 2,300; and total recoverable zinc, 3,700.

Goldschmid, J. "Water-Quality Aspects of Ground-Water Recharge in Israel." *Journal of the American Water Works Association.* **V. 66, n. 3. pp. 163–166. 1974.**

Because of the difference in rainfall and water needs in northern and southern Israel, a method of recharging groundwater and transporting it from the north to the south was needed. This article details the new system, including problems encountered and overcome.

Goolsby, D.A. *Geochemical Effects and Movement of Injected Industrial Waste in a Limestone Aquifer.* **Memoir No. 18. American Association of Petroleum Geologists. 1972.**

Since 1963, more than 6 billion gal of acidic industrial waste has been injected into a limestone aquifer near Pensacola, Florida. The industrial waste, an aqueous solution containing nitric acid, inorganic salts, and numerous organic compounds, is injected through two wells into the aquifer between depths of 1,400 and 1,700 ft (425-520 m). The aquifer receiving the waste is overlain by an extensive clay confining layer which, at the injection site, is about 200 ft (60 m) thick.

Industrial waste is presently (late 1971) being injected at a rate of about 2,100 gal per minute. Wellhead injection pressures are about 175 psi. Calculations indicate that pressure effects in the receiving aquifer extend out more than 30 mi (48 km). No apparent change in pressure has been detected in the aquifer directly above the clay confining layer. Geochemical effects were detected at a monitor well in the receiving aquifer 0.25 mi (0.4 km) from the injection wells about 10 months after injection began. The geochemical effects included increases in calcium-ion concentration and total alkalinity and formation of large quantities of nitrogen and methane gas.

Geochemical effects have not been detected at monitor wells in the receiving aquifer 1.9 mi (3.0 km) north and 1.5 mi (2.4 km) south of the injection wells, nor have effects been detected in a monitor well at the injection site open to the aquifer directly overlying the clay confining layer. Tests made at the injection wells early in 1968 indicated that rapid denitrification and neutralization of the waste occurred near the injection wells. Denitrification may have accounted for more than half neutralization, and solution of calcium carbonate accounted for the rest. Denitrification has not been observed since mid-1968, when the pH of the injected waste was lowered from 5.5 to 3.

Greene, G.E. *Ozone Disinfection and Treatment of Urban Storm Drain Dry-Weather Flows: A Pilot Treatment Plant Demonstration Project on the Kenter Canyon Storm Drain System in Santa Monica.* **The Santa Monica Bay Restoration Project, Monterey Park, CA. 1992.**

The Pico-Kenter Canyon storm drain has become the archetype for assessing the problems and possible solutions that can be associated with many of the urban storm drains in the Santa Monica Bay region. While known events of chemical contamination are few, the drain has long been known to be contaminated with indicator bacteria such as total and fecal coliforms. More recently, the consistent identification of human enteric viruses, f-male specific coliphage, and high densities of enterococcus bacteria have indicated that a potentially serious public health threat exists. The City of Santa Monica, with the assistance of the Santa Monica Bay Restoration Project (SMBRP), the United States Environmental Protection Agency (EPA), and the UCLA Laboratory of Biomedical and Environmental Sciences (LBES), recently completed an evaluation of ozone for the treatment of dry-weather storm drain flows. The primary goals of this study were to establish if ozone could be used to disinfect the water that typically flows from the Pico-Kenter storm drain and determine if some known hazardous chemical contaminants were present at significant levels.

Recently, ozone has become renowned in the drinking water industry as an alternative to chlorine that rapidly disinfects water while forming few halogenated byproducts. This study demonstrated that ozone was an effective disinfectant, reducing bacterial and viral populations by 3–5 log (99.9 to 99.999% of the microbes killed or inactivated). In many of the 438 effluent samples, coliform concentrations were sufficiently reduced to qualify the water for reclamation projects such as landscape irrigation along the Santa Monica Freeway, suggesting a possible useful role for the treated effluent. Ozonation byproducts (aldehydes) were detected in the plant effluent at low (<100 ppb) concentrations. No significant increase in halogenated byproducts, or mutagenicity, were observed following ozone disinfection. During a test of the ozonation process, twelve organic chemicals were added to the influent water and the effluent monitored. While some refractory compounds passed through the pilot facility intact, the concentrations of most were reduced.

In comparison to the State Ocean Plan Water Quality Objectives and Federal Drinking Water Maximum Contaminant Levels, the primary hazardous chemical constituents in the influent storm drain water were metals (primarily copper and lead) and polynuclear aromatic hydrocarbons (PAHs). While lead levels were significantly above both standards, the concentration of copper was well under drinking water standards. The mean observed levels of six major PAHs were approximately equal to their proposed phase V drinking water MCL standard (100–400 ng/L or pptr). Isolated samples were found to contain organic contaminants, such as ortho-xylene and the pesticide chlordane. This did not appear to be a pervasive problem, and can be attributed to isolated events that cannot be anticipated and will only be prevented through an informed and concerned public.

While the metal content of the water cannot be reduced using ozone, this study found that high concentrations of some organics, including PAHs, can be reduced during the ozonation process. This remediation probably occurs by oxidation and hydroxylation to less hazardous forms. Regardless of further ozonation investigations, additional more sensitive and definitive PAH analyses are warranted in future studies of the storm drain water and sediments.

Based on the results of this investigation, the City of Santa Monica is investigating construction of a disinfection facility that would reclaim high quality water for landscape irrigation, use low quality water for sewer flushing, and disinfect the remainder prior to releasing it into the Santa Monica Bay. Construction of the proposed facility would be encouraged by the support of the Santa Monica Bay Restoration Project in goal definition and consensus building among the member and nonmember agencies.

Summary conclusions:

1) Ozone at moderate doses (10–20 mg/L) was an extremely effective disinfectant of dry-weather storm drain flows.
2) Bacterial and viral levels were reduced 3–5 log (99.9 to 99.999% of the microbes killed or deactivated).
3) Much of the effluent was sufficiently disinfected to meet the landscape irrigation standard of 23 coliforms per 100 mL.
4) Based on California Ocean Plan Water Quality Objectives, heavy metals and polynuclear aromatic hydrocarbons appear to be the primary contaminants of concern in the pilot plant effluent.
5) While ozone disinfection byproducts were detected (aldehydes), their concentration was low, and in contrast to what would be expected from disinfection by chlorination, no increase in mutagenicity was observed following ozonation.

Summary recommendations:

1) The SMBRP should encourage further evaluation of the ozone disinfection process, by promoting the City of Santa Monica in its effort to design and construct a full-scale facility.
2) Since construction and operation of the proposed facility will require interagency consent and permitting, the City of Santa Monica solicits the continued assistance of the SMBRP in consensus building, policy direction, and technical support.
3) Further investigations into the use of the ozone technology should include provisions for the evaluation of Advanced Oxidation Processes (AOPs), using hydrogen peroxide and ozone, for the control of organic pollutants such as PAHs.

Hampson, P.S. *Effects of Detention on Water Quality of Two Stormwater Detention Ponds Receiving Highway Surface Runoff in Jacksonville, Florida.* **Geological Survey Water-Resources Investigations Report 86-4151. U.S. Geological Survey, Tallahassee, Florida. 1986.**

Water and sediment samples were analyzed for major chemical constituents, nutrients, and heavy metals following ten storm events at stormwater detention ponds that receive highway surface runoff in the Jacksonville, Florida, metropolitan area. The purpose of the sampling program was to detect changes in constituent concentration with time of detention within the pond system. Statistical inference of a relation with total rainfall was found on the initial concentrations of 11 constituents and with antecedent dry period for the initial concentrations of three constituents. Based on graphical examination and factor analysis, constituent behavior with time could be grouped into five relatively independent processes for one of the ponds. The processes were (1) interaction with shallow groundwater systems, (2) solubilization of bottom materials, (3) nutrient uptake, (4) seasonal changes in precipitation, and (5) sedimentation. Most of the observed water quality changes in the ponds were virtually complete within three days following the storm event.

Harper, H.H. *Effects of Stormwater Management Systems on Groundwater Quality.* **DER Project WM190. Florida Department of Environmental Regulation, Orlando, Florida. 1988.**

It has long been recognized that nonpoint sources of pollution contribute significantly to receiving water loadings of both nutrients and toxic elements such as heavy metals (Harper 1983; Sartor, et al. 1974). As a means of protecting Florida surface waters from the effects of nonpoint source pollution, the Florida Department of Environmental Regulation has established regulations that require new developments or projects to retain or detain specified volumes of runoff water onsite. In most cases runoff is collected in shallow ponds, which infiltrate all or part of the retained or detained volumes into groundwaters.

When stormwater management facilities receive inputs of stormwater containing nutrients, heavy metals and other pollutants, processes such as precipitation, coagulation, settling, and biological uptake deposit a large percentage of the input mass into the sediments. Recently, concern has been expressed that this continual accumulation of pollutants in the sediments of stormwater management ponds may begin to present a toxicity or pollution potential to underlying groundwaters. Specifically, do these pollutant accumulations cause physical and chemical changes to occur within the sediments of stormwater management facilities which mobilize certain pollutant species from the sediment phase into the water phase?

Hickey, J.J. "Subsurface Injection of Treated Sewage into a Saline-Water Aquifer at St. Petersburg, Florida—Aquifer Pressure Buildup." *Ground Water.* **V. 22, n. 1. pp. 48–55. 1984.**

The city of St. Petersburg has been testing subsurface injection of treated sewage into the Floridan aquifer as a means of eliminating discharge of sewage to surface waters and as a means of storing treated sewage for future nonpotable reuse. The injection zone originally contained native saline groundwater that was similar in composition to seawater. The zone has a transmissivity of about 1.2×10^6 ft^2/d and is within the lower part of the Floridan aquifer.

Treated sewage that had a mean chloride concentration of 170 mg/L was injected through a single well for 12 months at a mean rate of 4.7×10^5 ft^3/d. The volume of water injected during the year was 1.7×10^8 ft^3.

Pressure buildup at the end of one year ranged from less than 0.1 to as much as 2.4 psi in observation wells at the site. Pressure buildup in wells open to the upper part of the injection zone was related to buoyant lift acting on the mixed water in the injection zone, in addition to subsurface injection through the injection well.

Calculations of the vertical component of pore velocity in the semiconfining bed underlying the shallowest permeable zone of the Floridan aquifer indicate upward movement of native water. This is consistent with the 200- to 600-mg/L increase in chloride concentration observed in water from the shallowest permeable zone during the test.

Hickey, J.J. and J. Vecchioli. *Subsurface Injection of Liquid Waste with Emphasis on Injection Practices in Florida.* **Geological Survey Water-Supply Paper 2281. United States Government Printing Office, Washington D.C. 1986.**

Subsurface injection of liquid waste is used as a disposal method in many parts of the country. It is used particularly when other methods for managing liquid waste are either not possible or too costly. Interest in subsurface injection as a waste disposal method stems partly from recognition that surface disposal of liquid waste may establish a potential for degrading freshwater resources. Where hydrogeologic conditions are suitable and where surface disposal may cause contamination, subsurface injection is considered an attractive alternative for waste disposal. Decisions to use subsurface injection need to be made with care, because where hydrogeologic conditions are not suitable for injection, the risk to water resources, particularly groundwater, could be great. Selection of subsurface injection as a waste disposal method requires thoughtful deliberation and, in some instances, extensive data collection and analysis.

Subsurface injection is a good geological method of waste disposal. Therefore, many state and local governmental officials and environmentally concerned citizens who make decisions about waste disposal alternatives may know little about it. This report serves as an elementary guide to subsurface injection and presents subsurface injection practices in Florida as an example of how one state is managing injection.

Hickey, J.J. and W.E. Wilson. *Results of Deep-Well Injection Testing at Mulberry, Florida.* **Geological Survey Water-Resources Investigations Report 81-75. U.S. Geological Survey, Tallahassee, Florida. 1982.**

At the Kaiser Aluminum and Chemical Corporation plant, Mulberry, Florida, high-chloride, acidic liquid wastes are injected into a dolomite section at depths below about 4,000 ft below land surface. Sonar caliper logs made in April 1976 revealed a solution chamber that is about 100 ft in height and has a maximum diameter of 23 ft in the injection zone.

Results of two injection tests in 1972 were inconclusive because of complex conditions and the lack of an observation well that was open to the injection zone. In 1975, a satellite monitor well was drilled 2,291 ft from the injection well and completed open to the injection zone. In April 1975 and September 1976, a series of three injection tests were performed. Duration of the tests ranged from 240 to 10,020 minutes and injection rates ranged from 110 to 230 gpm. Based on an evaluation of the factors that affect hydraulic response, water level data suitable for interpretation of hydraulic characteristics of the injection zone were identified to occur from 200 to 1,000 minutes during the 10,020-minute test. Test results indicate that leakage through confining beds is occurring.

Transmissivity of the injection zone was computed to be within the range from 700 to 1,000 ft^2 day and the storage coefficient of the injection zone was computed to be within the range from 4×10^{-5} to 6×10^{-5}. The confining bed accepting most of the leakage appears to be the underlying bed. Also, it appears that the overlying beds are probably relatively impermeable and significantly retard the vertical movement of neutralized waste effluent.

Higgins, A.J. "Impacts on Groundwater Due to Land Application of Sewage Sludge." *Water Resources Bulletin.* **V. 20, n. 3. pp. 425–434. 1984.**

The project was designed to demonstrate the potential benefits of utilizing sewage sludge as a soil conditioner and fertilizer on Sassafras sandy loam soil. Aerobically digested liquid sewage sludge was applied to the soil at rates of 0, 22.4 and 44.8 Mg of dry solids/ha for three consecutive years between 1978 and 1981. Groundwater, soil, and crop contamination levels were monitored to establish the maximum sewage solids loading rate that could be applied without causing environmental deterioration. The results indicate that 22.4 Mg of dry solids/ha of sludge is the upper limit to ensure protection of the groundwater quality on the site studied. Application rates at or slightly below 22.4 Mg of dry solids/ha are sufficient for providing plant nutrients for the dent corn and rye cropping system utilized in the study.

Horsley, S.W. and J.A. Moser. "Monitoring Ground Water for Pesticides at a Golf Course—A Case Study on Cape Cod, Massachusetts." *Ground Water Monitoring Review.* **V. 10. pp. 101–108. 1990.**

The town of Yarmouth, Massachusetts, proposed to locate a new municipal golf course within a delineated area of recharge to public water supply wells. Two concerns of town officials were (1) hydrologic impacts upon downgradient wells; and (2) water quality impacts from fertilizers and pesticides. In response to these concerns, a thorough hydrogeologic investigation was made, fertilizer and pesticides management programs were recommended, and a groundwater monitoring program was developed.

The golf course parcel was determined to be underlain by a sand and gravel aquifer composed primarily of glacial outwash. Water table maps confirmed that groundwater flow was in the direction of several public water supply wells. A three-dimensional finite-difference flow model was used to determine the optimal location and pumping rates for irrigation wells. Potential nitrate-nitrogen concentrations in the groundwater were predicted to range from 5.0 to 7.9 mg/L, so slow-release fertilizers were recommended.

With the assistance of the EPA Office of Pesticide Programs, the list of proposed pesticides was reviewed and sorted into three categories based on the known leachability, mobility, and toxicity characteristics of each compound. Specific recommendations were made as to pesticide selection and application rate using that classification.

A monitoring program was developed to provide an ongoing assessment of any effects on water quality related to the application of fertilizer or pesticide. The elements of the monitoring program include (1) specifications for monitoring wells and lysimeters, (2) a schedule for sampling and analysis, (3) specific concentrations of nitrates or pesticide compounds that require resampling and analysis, restriction of usage, or remedial action, and (4) regular reports to the Yarmouth Water Quality Advisory Committee and to the Yarmouth Water Department. In an effort to ensure the implementation of this program, a table of responsibilities was prepared, and a Memorandum of Understanding adopting the program was signed by the town agencies interested in water supply protection and the golf course operation.

The monitoring facilities were installed with minimal problems as part of the golf course construction tasks. However, implementation of the sampling and analysis part of the program was accomplished only after some difficulty and delay. The assistance of the State Pesticide Bureau, the University of Massachusetts Department of Entymology, and the Massachusetts Pesticide Laboratory was enlisted when budgetary problems threatened to prevent implementation. It is apparent from Yarmouth's experience that the mere preparation of a plan is not sufficient by itself. Consultants who prepare the plan should make every possible effort to include implementation in their scope of services.

Hull, R.W. and M.C. Yurewicz. *Quality of Storm Runoff to Drainage Wells in Live Oak, Florida, April 4, 1979.* **Geological Survey Open-File Report 79-1073. U.S. Government Printing Office, Washington, D.C. 1979.**

Water quality samples of storm runoff to drainage wells were collected during a storm event on April 4, 1979. Two sites in commercial areas and two in residential areas of Live Oak, Florida, were sampled. A composite rainfall sample was collected from these sites, and rainfall quantity data were obtained from two additional sites.

Samples of storm runoff were analyzed for those constituents important to the potability of water. The analyses generally included filtered and unfiltered nutrients, bacteria, trace elements, and organics. Several of the analyses had constituent values that equaled or exceeded maximum containment levels for state primary drinking water standards and federal proposed secondary drinking water standards.

Ishizaki, K. "Control of Surface Runoff by Subsurface Infiltration of Stormwater: A Case Study in Japan." *Artificial Recharge of Groundwater.* **Butterworth Publishers, Boston. pp. 565–575. 1985.**

Even though the annual rainfall in Japan is about 1,800 mm, which is twice as much as the average rainfall worldwide, water shortages often occur during the summer months. This is because a large amount of surface water is diverted to flood rice fields, and as a result, river flow is reduced to the minimum.

There have been various long-term national programs for the development and management of water resources. Groundwater provides approximately 16% of the water resources requirements in Japan. In some areas, however, the groundwater levels have been lowered by excessive pumping, and such problems as land subsidence and seawater intrusion have occurred. There have been many cases, particularly in the coastal regions, where salinity of groundwater has become too high for any significant beneficial uses. In addition, due to the urbanization of river drainage basins in many parts of Japan, the decrease of soil infiltration has caused excessive surface runoff, leading to several serious flooding incidences in the surrounding areas.

To reduce excessive surface runoff and to promote groundwater recharge of stormwater, a number of groundwater recharge methods have been investigated at the Japan Ministry of Construction's Public Works Research Institute. This report describes the subsurface infiltration of stormwater through culverts and also discusses the effects of the stormwater infiltration experiment at an apartment complex in Tokyo, Japan.

Jansons, J., L.W. Edmonds, B. Speight and M.R. Bucens. "Movement of Viruses after Artificial Recharge." *Water Research*. V. 23, n. 3. pp. 293–299. 1989a.

Results of human enteric virus movement through soil and groundwater aquifers after artificial recharge using wastewater are presented. The penetration through the recharge soil of indigenous viruses from treatment plant effluent was found to be much greater than that of a seeded vaccine poliovirus. Echovirus type 11 from wastewater was detected at a depth of 9.0 m in groundwater from a bore located 14 m from the recharge basin, whereas seed poliovirus was not isolated beyond a depth of 1.5 m below the recharge basin. It was concluded that data on enteric virus survival in groundwater were required if safe abstraction distances were to be determined.

Jansons, J., L.W. Edmonds, B. Speight and M.R. Bucens. "Survival of Viruses in Groundwater." *Water Research*. V. 23, n. 3. pp. 301–306. 1989b.

Survival in groundwater of echovirus types 6, 11, and 24, coxsackievirus type B5 and poliovirus type 1 was determined. Enterovirus survival in groundwater was found to be variable and appeared to be influenced by a number of factors: temperature, dissolved oxygen concentration, and possibly the presence of microorganisms. Dissolved oxygen concentration was the most significant factor in loss of virus in groundwater. Poliovirus type 1 incubated in groundwater with a mean dissolved oxygen concentration of 2.0 mg/L decreased in infectivity by 100-fold in 50 days compared with 20 days for a decrease of the same magnitude when incubated in groundwater with a mean dissolved oxygen concentration of 5.4 mg/L. Echovirus type 6 was found to be least stable, and poliovirus type 1 was found to be most stable, although virus stability may have been due to conditions existing in individual groundwater bores.

Johnson, R.B. "The Reclaimed Water Delivery System and Reuse Program for Tucson, Arizona." *Irrigation Systems for the 21st Century, Portland, Oregon, 1987*. (American Society of Civil Engineers, New York, New York, pp. 344–351). 1987.

The City of Tucson has implemented a reclaimed water reuse program in a community-wide effort to preserve high-quality groundwater for potable and other priority uses. Presently, the system consists of an 8.2-million-gallons-per-day (mgd) pressure filtration plant (expandable to 25 mgd), a 3-million-gallon reservoir, a 12-mgd booster facility, and a significant network of large-capacity distribution lines that supply major turf irrigation uses throughout the community. An additional element of the system is a 1.0-mgd aquifer recharge facility that provides cost-effective seasonal storage of reclaimed water for subsequent recovery and use during the peak demand season.

The initial system was made operational in February, 1984, with the first deliveries of reclaimed water for turf irrigation. By the end of fiscal year 1986–87, Tucson Water will be providing nearly 5,000 ac-ft/yr of reclaimed water with the expanding reclaimed water system. It is projected that by 1995, Tucson's reclaimed water delivery system will be serving 35,000 ac-ft for turf irrigation, cooling tower, and gravel washing uses throughout the metropolitan area. The program has reduced peak demands on the potable water system, and represents a major step toward the efficient management of the water resources available to our growing community.

Karkal, S.S. and D.L. Stringfield. "Wastewater Reclamation and Small Communities: A Case History." *Water Environment Federation 65th Annual Conference & Exposition, New Orleans, Louisiana, 1992.* **(Water Environment Federation, Alexandria, Virginia, pp. 419–425). 1992.**

In California's Central Valley, regulatory requirements and concerns of the local community have led to a unique wastewater reclamation project. The City of Orange Cove owns and operates a 3,785.4 m³ (1.0 mgd) treatment plant that uses tertiary treatment to produce a high-quality reclaimed water that meets the California Code of Regulations (CCR) Title 22 requirements for unrestricted irrigation use. The water is supplied to the independently owned and operated Orange Cove Irrigation District (OCID). This facility is in stark contrast to many reclaimed wastewater irrigation facilities in the Central Valley that use primary or secondary effluents for irrigation. The objective of this paper is to discuss the regulatory requirements leading to this project, and the factors involved in the selection of the treatment processes used to meet these requirements.

Katopodes, N.D. and J.H. Tang. "Self-Adaptive Control of Surface Irrigation Advance." *Journal of Irrigation and Drainage Engineering.* **V. 116, n. 5. pp. 696–713. 1990.**

The controllability of surface irrigation is examined by analytical means and numerical tests based on the linearized zero-inertia model. First, the inflow hydrograph is proved to be identifiable from advance data. Then, it is shown that it is possible to control the advance rate by adjustment of the inflow rate. Field parameter heterogeneities are automatically taken into account, so a predetermined advance trajectory is obtained under arbitrary field conditions. The model utilizes a tentative time increment, during which a trial value for inflow is adopted. The resulting wave advance is simulated by the zero-inertia model. Discrepancies between the actual and desired advance rate are then used to construct an objective function, whose minimization leads to a correction of the inflow rate for the next time increment of inflow. Finally, examples of self-adaptive control are presented, in which an irrigation stream is led into a field of unknown parameters. The model uses real-time information for the identification of the parameters and simultaneous control of the inflow rate to achieve a desired advance rate.

Kaufman, M.I. "Subsurface Wastewater Injection, Florida." *Journal of Irrigation and Drainage Engineering.* **V. 99, n. 1. pp. 53–70. 1973.**

The Secretary of the Interior directed the U.S. Geological Survey in December, 1969 to begin a research program to evaluate the "effects of underground waste disposal on the Nation's subsurface environment, with particular attention to ground-water supplies." The directive noted the complexity of the subsurface environment and emphasized the need to begin collecting pertinent environmental data.

As a direct result, the Survey is engaged in an investigative and research program to develop a scientific basis for assessing the long-term environmental impact of subsurface waste injection. Prediction of movement, chemical interaction, and ultimate fate of injected liquid waste is difficult. As noted by Piper: "Uncritical acceptance [of deep-well injection] would be ill advised." The complexity of both the waste and the subsurface environment preclude making generalizations; with the present state of knowledge, a thorough regional and localized study must be made for each proposed waste-injection system.

Extensive areas in Florida are underlain by deep permeable saline-aquifer systems that are separated from overlying freshwater aquifers by low-permeability confining materials consisting of clay, evaporites, or dense carbonate rocks.

To resolve problems of waste disposal and to alleviate deterioration of fresh and estuarine waters, approaches such as deep-well injection of industrial and municipal effluents into these saline-aquifer systems are being actively explored.

The hydrologic and geochemical characteristics of these saline-aquifer systems and their response to waste injection are the subject of current research by the Survey. This information would assist planning-management and regulatory agencies in their evaluation of subsurface injection of liquid wastes, including its potential applicability to regional water and waste management systems.

This paper contains both a summary of data and present status of subsurface waste injection in Florida, including observed hydraulic and geochemical effects and a descriptive regional portrayal of the lithology and hydrogeochemistry of the saline-aquifer system.

Knisel, W.G. and R.A. Leonard. "Irrigation Impact on Groundwater: Model Study in Humid Region," *Journal of Irrigation and Drainage Engineering.* **V. 115, n. 5. pp. 823–839. 1989.**

The Groundwater Loading Effects of Agricultural Management Systems (GLEAMS) model was applied to estimate the effects of: (1) soil; (2) planting date; (3) irrigation level; and (4) pesticide characteristic on pesticide leaching below the root zone of representative coarse-grained soils. Climate/application/pesticide-characteristic interactions are shown to significantly affect pesticide losses, whereas irrigation practice has little effect. Persistent and mobile compounds exhibit the highest losses.

Krawchuk, B.P. and B.G.R. Webster. "Movement of Pesticides to Ground Water in an Irrigated Soil." *Water Pollution Research Journal of Canada.* **V. 22, n. 1. pp. 129–146. 1987.**

The movement of pesticide residues to groundwater was studied on a commercial farm southwest of Portage la Prairie, Manitoba. The site had sandy soil with low organic matter content, a high water table, a tile drain system and an irrigation system using river water. Records were available from the beginning of commercial operation in 1979 describing pesticide usage on a field by field basis. A total of 21 different pesticide formulations were used in the 5 years of operation.

An initial (1981) random sampling of the tile drain water did not detect any pesticide residues in the outflow at the 0.02 µg/L level. A subsequent extensive sampling (1982) detected residues of chlorothalonil on eight occasions ranging from 0.06 to 3.66 mg/L in the tile drain outflow. Groundwater from one of two wells in the northwest quarter was found to contain chlorothalonil at a level of 10.1 to 272.2 mg/L in 1982 and 0.4 to 9.0 mg/L in 1983, carbofuran at a level of 11.5 to 158.4 mg/L in 1982 and <0.5 to 1.0 mg/L in 1983, and carbofuran phenol (not quantified) in 1982 and 1983.

RP-HPLC K_{ow} data indicated that a number of the pesticides used on the farm could be as mobile or more mobile than chlorothalonil, which had been detected in the groundwater in two consecutive years; however, of the other pesticides only carbofuran was detected in the groundwater. With a K_{ow} lower than that of chlorothalonil, carbofuran was expected to be more mobile than chlorothalonil, and to appear in the water sooner, but this was not observed in the field samples.

Ku, H.F.H., N.W. Hagelin and H.T. Buxton. "Effects of Urban Storm-Runoff Control on Ground-Water Recharge in Nassau County, New York." *Ground Water*. V. 30, n. 4. pp. 507–513. 1992.

Before urban development, most groundwater recharge on Long Island, New York, occurred during the dormant season, when evapotranspiration is low. The use of recharge basins for collection and disposal of urban storm runoff in Nassau County has enabled groundwater recharge to occur also during the growing season. In contrast, the use of storm sewers to route storm runoff to streams and coastal waters has resulted in a decrease in groundwater recharge during the dormant season. The net result of these two forms of urban storm runoff control has been an increase in annual recharge of about 12% in areas served by recharge basins and a decrease of about 10% in areas where runoff is routed to streams and tidewater. On a countrywide basis, annual groundwater recharge has remained nearly the same as under predeveloped conditions, but its distribution pattern has changed. Redistribution resulted in increased recharge in the eastern and central parts of the county, and decreased recharge in the western and near shore areas. Model simulation of recharge indicates that the water table altitude has increased by as much as 5 ft above predevelopment levels in areas served by recharge basins and declined by as much as 3 ft in areas where stormwater is discharged to streams and tidewater.

Ku, H.F.H. and D.L. Simmons. *Effect of Urban Stormwater Runoff on Ground Water Beneath Recharge Basins on Long Island, New York*. **Geological Survey Water-Resources Investigations Report 85-4088. U.S. Geological Survey, Syosset, New York. 1986.**

Urban stormwater runoff was monitored during 1980–82 to investigate the source, type, quantity, and fate of contaminants routed to the more than 3,000 recharge basins on Long Island and to determine whether this runoff might be a significant source of contamination to the groundwater reservoir. Forty-six storms were monitored at five recharge basins in representative land use areas (strip commercial, shopping mall parking lot, major highway, low-density residential, medium-density residential).

Runoff/precipitation ratios indicate that all storm runoff is derived from precipitation on impervious surfaces in the drainage area except during storms of high intensity or long duration, when additional runoff can be derived from precipitation on permeable surfaces.

Concentrations of most measured constituents in individual stormwater samples were within federal and state drinking water standards. The few exceptions are related to specific land uses and seasonal effects.Lead was present in highway runoff in concentrations up to 3,300 µg/L, and chloride was found in parking lot runoff in concentrations up to 1,100 mg/L during winter, when salt is used for deicing.

The load of heavy metals was largely removed during movement through the unsaturated zone, but chloride was not removed. Total nitrogen was commonly found in greater concentrations in groundwater than in stormwater; this is attributed to seepage from cesspools and septic tanks and to the use of lawn fertilizers.

In the five composite stormwater samples and nine groundwater grab samples that were analyzed for 113 U.S. Environmental Protection Agency–designated "priority pollutants," four constituents were detected in concentrations exceeding New York State guidelines of 50 µg/L for an individual organic compound in drinking water: p-chlorom-cresol (79 µg/L in groundwater at the highway basin); 2,4-dimethylphenol (96 µg/L in groundwater at the highway basin); 4-nitrophenol (58 µg/L in groundwater at the parking lot basin); and methylene chloride (230 µg/L in stormwater at the highway basin). One stormwater sample and two groundwater samples exceeded New York State guidelines for total organic compounds in drinking water (100 µg/L). The presence of these constituents is attributed to contamination from point sources rather than to quality of runoff from urban areas.

The median number of indicator bacteria in stormwater ranged from 10^8 to 10^{10} MPN/ 100 mL (most probable number per 100 mL). Fecal coliforms and fecal streptococci increased by 1 to 2 orders of magnitude during the warm season. Total coliforms concentrations showed no significant seasonal differences.

Low-density residential and nonresidential (highway and parking lot) areas contributed the fewest bacteria to stormwater; medium-density residential and strip commercial areas contributed the most. No bacteria were detected in the groundwater beneath any of the recharge basins.

The use of recharge basins to dispose of storm runoff does not appear to have significant adverse effects on groundwater quality in terms of the chemical and microbiological stormwater constituents studied.

Ku, H.F.H., J. Vecchioli and S.E. Ragone. "Changes in Concentration of Certain Constituents of Treated Waste Water During Movement through the Magothy Aquifer, Bay Park, New York." *Journal Research U.S. Geology Survey*. V. 3, n. 1. pp. 89– 92. 1975.

Approximately 7 million gallons (27 million liters) of tertiary-treated sewage (reclaimed water) was injected by well into the Magothy aquifer and was subsequently pumped out. As the reclaimed water moved through the aquifer, concentrations of certain dissolved constituents decreased as follows: total nitrogen, 7%; methylene blue active substances, 49%; chemical oxygen demand, 50%; and phosphate, more than 93%.

Lauer, D.A. "Vertical Distribution in Soil of Sprinkler-Applied Phosphorus." *Soil Science Society of America Journal.* **V. 52, n. 3. pp. 862–868. 1988a.**

Soil-immobile plant nutrients, such as phosphorus, accumulate near the soil surface in conservation cropping systems where tillage leaves crop residues on or near the soil surface and limits soil mixing. The objective was to determine, in field and laboratory experiments, the vertical distribution in soil of phosphorus applied in sprinkler irrigation water. Following phosphorus applications, distribution was determined from depth increment (1 or 2 cm) sampling on a Warden silt loam (coarse-silty, mixed, mesic, xerollic camborthids) and a Quincy sand (mixed, mesic, xeric torripsamments). The fertilizer materials applied in 10-mm irrigations were: monoammonium phosphate (227 g P kg^{-1}), urea phosphate (192 g P kg^{-1}), commercial white phosphoric acid (238 g P kg^{-1}), and ammonium polyphosphate (149 g P kg^{-1}) containing 61g P kg^{-1} as polyphosphate. Application rates ranged from 50 to 400 kg P ha^{-1}. Movement of P was about the same for all P fertilizer materials except ammonium polyphosphate, from which P moved only 60 to 70% of the depth of the other materials. Postapplication irrigation totals up to 160 mm at 10 m d^{-1}, which was applied without drying cycles, distributed P more uniformly with depth. The overall mean depths of P penetration across all treatments were 10.4 cm (s.d. = 4.04 cm) on the Quincy sand and 7.3 cm (s.d. = 4.93 cm) on the Warden silt loam. These depths of penetration and vertical distribution of sprinkler-applied P are probably sufficient to supply the P needs of crop plants under sprinkler irrigation.

Lauer, D.A. "Vertical Distribution in Soil of Unincorporated Surface-Applied Phosphorus under Sprinkler Irrigation." *Soil Science Society of America Journal.* **V. 52, n. 6. pp. 1685–1692. 1988b.**

Determining vertical distribution of phosphorus is important in irrigated conservation cropping systems for evaluation of phosphorus fertilization because soil-immobile phosphorus accumulates near the soil surface where limited tillage reduces soil mixing. The objective was to determine in field and laboratory experiments the vertical distribution of phosphorus from surface-applied monoammonium phosphate (MAP; 227 g P kg^{-1}), triple superphosphate (TSP; 197 g P kg^{-1}) and ammonium polyphosphate (APP; 149 g P kg^{-1}). Following phosphorus application, vertical distribution was determined from 2-cm-depth increment samples in a Quincy sand (mixed, mesic, xeric torripsamments), a Warden silt loam (coarse-silty, mixed, xerollic camborthids), and a calcareous subsoil of the Warden. There was little effect on phosphorus distribution from antecedent moisture; fertilizer rates at 30, 60, 120, or 240 kg P ha^{-1}; or from preirrigation reaction times of 1, 4, or 16 day. Continuous postapplication irrigation totals of 40 or 160 mm at 10 mm day^{-1} moved phosphorus somewhat deeper into the soil, principally at 160 mm on the Quincy sand. Overall mean penetration depths of phosphorus were as follows: (i) APP moved the farthest in the Quincy sand (mean = 6.1; s.d. = 1.05 cm); (ii) penetrations were practically the same for MAP or TSP on the Quincy sand (mean = 5.5; s.d. = 1.25 cm); (iii) depth of penetration was intermediate for APP on the noncalcareous Warden (mean = 4.0; s.d. = 0.98 cm); and (iv) downward movement in calcareous Warden of phosphorus from all fertilizer phosphorus materials was much more restricted (MAP/TSP: mean = 3.1; s.d. = 0.10 cm and APP: mean = 3.3; s.d. = 0.52) than on the two noncalcareous soils. Overall, the most apparent conclusion from this study is that the reactivity of phosphorus fertilizer material with the soil is the dominant and overriding determinant of the vertical distribution of surface-applied phosphorus.

Lee, E.W. "Drainage Water Treatment and Disposal Options." *Agricultural Salinity Assessment and Management.* **American Society of Civil Engineers. pp. 450–468. 1990.**

Treating and disposing of subsurface drainage water from irrigated agricultural lands presents unique technical challenges. A review of the literature reveals limited experiences in the management of such waters. The challenge is made more difficult by the complex chemical characteristics of most drainage water. Drainage usually contains a heavy salt load—a perennial cause for concern—and residual pesticides, herbicides, fungicides, fertilizers, and toxic trace elements—a more recent cause for concern. While crop management practices and conventional treatment processes can control salt and residuals to some degree, a new treatment technology needs to be developed to control toxic trace elements. Conventional methods appear to be ineffective in meeting the requirements set by many regulatory agencies.

In this report, the treatment and disposal of subsurface drainage from irrigated lands will be covered, disposal options will be presented, the technology of drainage water treatment and disposal options will be reviewed, and current research on treatment technology will be discussed.

Lloyd, J.W., D.N. Lerner, M.O. Rivett and M. Ford. "Quantity and Quality of Groundwater Beneath an Industrial Conurbation—Birmingham, UK." *Hydrological Processes and Water Management in Urban Areas, Duisburg, Federal Republic of Germany, 1988.* **(International Hydrological Programme, UNESCO, pp. 445–453). 1988.**

Since the 1960s groundwater heads have been rising noticeably beneath Birmingham, principally due to reducing abstraction. Despite the almost complete urbanization of the aquifer surface, potential recharge is at least as high as it was before the city was built. There is sufficient leakage from water mains and sewers to make up for the reduced infiltration at the surface. The marked change in groundwater conditions has prompted interest in groundwater quality, and both inorganic and organic quality of the groundwater are currently being studied. First indications are of widespread worsening inorganic quality, with high nitrates, chlorides, and certain trace elements. Chlorinated organic solvents (e.g. trichloroethylene) are widespread, but there is little evidence to date of other organic pollutants.

Loh, P.C., R.S. Fujioka and W.M. Hirano. "Thermal Inactivation of Human Enteric Viruses in Sewage Sludge and Virus Detection by Nitrose Cellulose-Enzyme Immunoassay." *Chemical and Biological Characterization of Sludges, Sediments, Dredge Spoils, and Drilling Muds. ASTM STP 976.* **American Society for Testing and Materials, Philadelphia, Pennsylvania. pp. 273–281. 1988.**

The Zimpro Thermal Sludge Treatment Process installed at the Sand Island Wastewater Treatment Plant, Honolulu, Hawaii, was evaluated for its reliability in disinfecting human enteric viruses and fecal bacteria in the treated sludge. The principle of this process involves grinding the sludge particles to a small size (<4.8 mm) and heating the ground sludge to 193°C under 330 psi pressure for 30 min. Such thermally treated sludge yielded no human enteric viruses and little or no fecal bacteria (<2 to 24 MPN/100 g), thus rendering the sludge safe for reuse. In corollary studies, the nitrocellulose-enzyme immunoassay was evaluated as an alternative cost-effective method to augment infectivity assays for the detection of human enteric viruses. The method was found to be rapid, highly sensitive (it can detect picogram quantities), and specific for the detection of human enteric viruses.

Malik, A., M. Stone, F.R. Martinez and R. Paul. "First Wastewater Desalting Plant in Central Coast, California." *Water Environment Federation 65th Annual Conference & Exposition, New Orleans, Louisiana, 1992.* **(Water Environment Federation, Alexandria, Virginia, pp. 395–406). 1992.**

This paper describes a 144.6 L/s (3.0 mgd) water reclamation plant and a 35 L/s (0.8 mgd) reverse osmosis (RO) plant. The water reclamation plant treats secondary treated discharge from the Goleta Sanitary District wastewater treatment plant. The RO plant further processes a portion of the effluent that will be produced at the water reclamation plant. RO treatment is needed to: (1) lower the chloride content of the tertiary effluent to less than 300 mg/L, for protection of golf course greens and sensitive plants; and (2) to provide water with reduced concentrations of chlorides, sodium, sulfates, and other constituents to agricultural growers.

The installed RO Plant will produce 35 L/s (0.8 mgd) of desalted water. The facility will discharge 9.6 L/s (0.22 mgd) desalted water to the agricultural growers and 25.4 L/s (0.58 mgd) to reclaimed water storage tanks for blending with effluent from the water reclamation plant. 115.7 L/s (2.64 mgd) of blended effluent will be available to distribute to all other reclaimed water users.

Mancini, J.L. and A.H. Plummer Jr. "A Method for Developing Wet Weather Water Quality Criteria for Toxics." *Water Environment Federation 65th Annual Conference & Exposition, New Orleans, Louisiana, 1992.* **(Water Environment Federation, Alexandria, Virginia, pp. 15–26). 1992.**

Current federal and state water quality programs are focusing on control of toxics and on impacts from diffuse sources. These two elements of regulatory programs can be expected to come together in the future. Regulations will begin to address the effects of toxics from diffuse sources.

Diffuse sources of contaminants, such as runoff from urban, industrial, and agricultural sites are associated with wet-weather events. The inputs of toxic materials from these sources are intermittent, and may not occur during all runoff events. Exposure of aquatic organisms to toxicants from these sources is often of short duration, separated by extended periods that provide opportunities for organism recovery. The concentrations and exposure patterns of toxicants that cause impacts on resident aquatic organisms are different for diffuse wet-weather dischargers and continuous point source inputs.

The potential cost for control of toxics in discharges from diffuse sources is large. Therefore, it is important that water quality criteria properly represent the level of protection needed to address contaminants introduced by diffuse sources. The application of technologically based modifications to existing water quality criteria that take into account the characteristics of diffuse sources (e.g. wet-weather inputs) could produce environmental protection at lower costs.

This paper presents technology which could be used to modify existing EPA numerical criteria for toxics to consider the effects of the variability of concentrations and exposure patterns associated with wet-weather inputs. Illustrations of criteria development, site-specific use of criteria, and compliance monitoring are included. The technology can also be used to estimate the impacts of wet-weather discharges on aquatic organisms.

Markwood, I.M. "Waterborne Disease—Historical Lesson." *Ground Water.* **V. 17, n. 2. pp. 197–198. 1979.**

While it is true that waterborne diseases are still with us, and probably always will be, we cannot classify them as a current threat in the sense that they were 100 years ago. The discovery that chlorine would disinfect water supplies removed these diseases from a "current threat" to a "historical lesson" category. We are not faced with unknown diseases that we are unable to attack. We have only to look at what others have done to protect themselves and follow the same or improved practices.

If the record of waterborne outbreaks in public water supplies in this country from the end of World War II up to the present is examined, it will be found that all are caused by breakdowns in disinfection procedures, or carelessness. The record is replete with statements such as "improper disinfection after repair," "breakdown or lack of disinfecting equipment," "back siphonage," and other similar statements all pointing to failure to follow practices which the history of water treatment has shown to be necessary for protection against waterborne disease. Carelessness allows recurrence of disease outbreaks. If the lessons of history were followed, the conquest of waterborne disease transmission by public water systems could be complete.

Marton J. and Mohler I. "The Influence of Urbanization on the Quality of Groundwater." *Hydrological Processes and Water Management in Urban Areas, Duisburg, Federal Republic of Germany, 1988.* **(International Hydrological Programme, UNESCO, pp. 452–461). 1988.**

In the example of the described anthropogenic activities in an urbanized basin we have manifested their influence upon a quantitative and qualitative regime of groundwater on the territory of Bratislava and the consequential measures taken to suppress or liquidate their negative effect as well.

Marzouk, Y., S.M. Goyal and C.P. Gerba. "Prevalence of Enteroviruses in Ground Water of Israel." *Ground Water.* **V. 17, n. 5. pp. 487–491. 1979.**

Few studies have been performed on the occurrence of enterovirus contamination of groundwater. In this study, 99 groundwater samples were examined for the presence of enteroviruses, total bacteria, fecal coliforms, and fecal streptococci by standard methods. Enteroviruses were isolated from 20% of the samples. Viruses were isolated from 12 samples that contained no detectable fecal organisms per 100 mL. No statistical correlation between presence of virus and bacteriological indicators could be determined. The widespread failure of current bacteriological standards to indicate the presence of potentially pathogenic enteroviruses in groundwater is an area of concern that requires more study.

Merkel, B., J. Grossmann and P. Udluft. "Effect of Urbanization on a Shallow Quaternary Aquifer." *Hydrological Processes and Water Management in Urban Areas, Duisburg, Federal Republic of Germany, 1988.* **(International Hydrological Programme, UNESCO, pp. 461–469). 1988.**

Quantity and quality of groundwater renewal is mainly determined by physico-chemical processes within the soil and the unsaturated zone. Instances of pollution of groundwater from urbanization were indicated by deposits from precipitation, sodium chloride spreading, and leakage from sewerage.

Mossbarger, W.A. Jr. and R.W. Yost. "Effects of Irrigated Agriculture on Groundwater Quality in Corn Belt and Lake States." *Journal of Irrigation and Drainage Engineering.* **V. 115, n. 5. pp. 773–789. 1990.**

The impact of irrigation on groundwater quality is influenced by climate, topography, geology, soils, geohydrology, crops, and agricultural practices. Since the early 1950s, the irrigated crop acreage in the Corn Belt and Lake States has increased markedly. Irrigation in these regions is concentrated in areas underlain by sandy soils with low moisture-holding capacities, where supplemental moisture and relatively heavy applications of agrichemicals are needed to achieve economically viable crop yields. Due to the high hydraulic conductivities and low attenuation capacities of sandy soils, shallow aquifers underlying these areas are particularly susceptible to contamination with nitrates and stable, soluble pesticides. Present and potential problems associated with irrigation in these states are illustrated by available case studies from the Central Sand Plain of Wisconsin.

Nellor, M.H., R.B. Baird and J.R. Smyth. "Health Aspects of Groundwater Recharge." *Artificial Recharge of Groundwater.* **Butterworth Publishers, Boston. pp. 329–355. 1985.**

Southern California, like many semiarid regions of the United States, does not receive sufficient water from local sources to support the considerable population of the area. Almost two-thirds of the water supply is imported 200 to 500 miles from the point of use. The remainder is derived from local groundwater basins. In some areas, the occurrence of overdraft conditions and saltwater intrusion has led to the adjudication of groundwater extractions and/or the implementation of artificial groundwater replenishment. Water sources used for groundwater replenishment include storm runoff, imported water, and, in some cases, treated wastewater (reclaimed water).

There is considerable uncertainty at this time regarding the sufficiency of water supplies for future water needs of the area. Population growth projections, coupled with reductions in imported water deliveries, indicate that by the mid-1990s water needs may exceed available supplies. These water shortage predictions have stimulated regional planning activities aimed at optimizing available water supplies through conservation efforts and development of new local sources of supply through conjunctive groundwater storage and water reclamation. Foremost among these planning efforts is the Orange and Los Angeles Counties Water Reuse Study, which has identified the most viable water reclamation projects within the South Coast Region and developed a financial and institutional scheme for their implementation. Of all the reclamation projects under consideration, groundwater recharge represents the largest and most economical use of reclaimed water.

Despite these economic incentives, implementation of proposed groundwater recharge projects is constrained by concerns over the potential health impacts of indirect reuse for potable purposes. Health issues associated with groundwater recharge include the acute and chronic effects of trace metals, minerals, pathogens, and organic compounds that, if present in reclaimed water, may ultimately become part of a potable water supply. Available information on existing groundwater recharge projects has never shown any evidence of impaired water quality or health. Yet it is recognized that this information is insufficient for rigorous evaluation of the possible long-term health implications associated with indirect potable reuse.

The existing groundwater recharge projects in Los Angeles and Orange Counties provided an opportunity to gather the data needed to evaluate the health significance of water reuse by groundwater recharge. Foremost among these is the Whittier Narrows groundwater recharge project located in the Montebello Forebay area of Los Angeles County, where planned replenishment using reclaimed water has been practiced since 1962. A work plan was developed by the Los Angeles County Sanitation Districts, which incorporated multidisciplinary research recommendations proposed by a "blue ribbon" panel of experts convened by the California State Water Resources Control Board, the Department of Water Resources, and the Department of Health Services. The work plan formed the basis for the Health Effects Study that formally began in November 1978 and was completed in March 1984.

Nightingale, H.I., J.E. Ayars, R.L. McCormick and D.C. Cehrs. "Leaky Acres Recharge Facility: A Ten-Year Evaluation." *Water Resources Bulletin*. V. 19, n. 3. pp. 429–437. 1983.

From 1971–1980, studies were conducted at Fresno, California, to identify and quantify, where possible, the soil and water chemistry, subsurface geologic, hydrologic, biologic, and operational factors that determine the long-term (10-year) effectiveness of basin type artificial groundwater recharge through alluvial soils. This paper updates previous findings and refers to publications that describe the geology beneath the basins and regional geology that determine the transmission and storage properties for local groundwater management and chemical quality enhancement. High-quality irrigation water from the Kings River was used for recharge. Construction and land costs for the present expanded facility—83 ha (205.2 ac) using three parcels of land—were $1,457,100. The 9-yr annual mean costs for only canal water, maintenance, and operation were $110.42/ha-m ($13.62/ac-ft) based on an average recharge rate of 1338 ha-m/yr (10,848 ac-ft/yr) at 86% facility efficiency. The measured end of season recharge rate averaged 14.97 ± 0.24 cm/day. The 10-yr mean actual recharge rate based on actual water delivered, total ponded area, and total days of recharge was 12.1 cm/day.

Nightingale, H.I. "Accumulation of As, Ni, Cu, and Pb in Retention and Recharge Basin Soils from Urban Runoff." *Water Resources Bulletin*. V. 23, n. 4. pp. 663–672. 1987a.

The accumulation of arsenic, nickel, copper, and lead in the soil profile was determined beneath five urban stormwater retention/recharge basins used by the Fresno Metropolitan Flood Control District, California. Soils were sampled from the surface to the first zone of saturation and compared with soils from an adjacent uncontaminated control site. These elements were found to be accumulating in the first few centimeters of basin soil and are important to the effectiveness of a specific best management practice, i.e., the retention and recharge of urban stormwater. Study basins in use since 1962, 1965, and 1969 had lead contents in the 0–2 cm soil depth interval of 570, 670, and 1400 mg Pb/kg soil, respectively. The median indigenous soil lead concentration was 4.6 mg/kg soil. The practice of removing excess flood runoff water from two basins by pumping apparently is a factor in reducing the accumulation rate of these elements in the surface soils of the basins.

Nightingale, H.I. "Water Quality Beneath Urban Runoff Management Basins." *Water Resources Bulletin*. V. 23, n. 2. pp. 197–205. 1987b.

The chemical impact of urban runoff water on water quality beneath five retention/recharge basins was investigated as part of the US EPA's Nationwide Urban Runoff Program in Fresno, California. Soil water percolating through alluvium soils and the groundwater at the top of the water table were sampled with ceramic/Teflon® vacuum water extractors of depths up to 26 m during the two-year investigation. Inorganic and organic pollutants are present in the runoff water delivered to the basins. No significant contamination of percolating soil water or groundwater underlying any of the five retention/recharge basins has occurred for constituents monitored in the study. The oldest basin was constructed in 1962. The concentration of selected trace elements in the groundwater samples was similar to the levels reported in the regional groundwater. None of the pesticides or other organic priority pollutants, for which water samples were analyzed, was detected, except diazinon, which was found in trace counts (0.3 µg/L or less) in only three soil water samples. These results are important to the continued conservation of stormwater and the development of a best management practice for stormwater management using retention/recharge basins in a semi-arid climate.

Nightingale, H.I. and W.C. Bianchi. "Ground-Water Chemical Quality Management by Artificial Recharge." *Ground Water*. V. 15, n. 1. pp. 15–22. 1977a.

The effectiveness of basin groundwater recharge at the Leaky Acres Facility in Fresno, California for improving the regional groundwater quality was studied as 65,815,000 m³ of high-quality surface water was recharged from 1971 through 1975. Observation wells at the facility showed some variability in chemical parameters associated with each recharge period. The long-term decrease in salinity could be described by decay curve fitted by regression analysis.

Without a special network of observation wells outside the facility, scientific evaluation of the enclave of recharged water is not possible. A practical evaluation of water quality changes is possible from producing water wells around the facility. However, the pumping well discharge time variations, well depth, aquifer sequence, and prior use of surrounding land must be considered, since all of these factors affect the pumped water quality and its seasonal variability. Recharge at Leaky Acres had noticeably decreased the groundwater salinity for a distance of up to 1.6 km in the regional groundwater movement.

Nightingale, H.I. and W.C. Bianchi. "Ground-Water Turbidity Resulting from Artificial Recharge." *Ground Water*. V. 15, n. 2. pp. 146–152. 1977b.

Turbid groundwater is rarely observed in domestic or public supply aquifers. At the Leaky Acres Recharge Facility at Fresno, California, water of low salinity (<50 μmhos/cm) and turbidity (<5 FTU, Formazin turbidity units) is recharged in the spreading basins. Six months after the start of the third (1973) recharge period, the groundwater salinity was decreased to about 100 μmhos/cm from the initial mean of 147 μmhos/cm and the groundwater became visibly turbid (>5 FTU). Two months later, some peripheral domestic wells also began to become turbid. After two more recharge periods (1974 and 1975), turbidity at 10 observation wells beneath Leaky Acres averaged 18 FTU and salinity averaged 74 μmhos/cm. By this time, groundwater turbidity in peripheral wells near Leaky Acres had decreased to <0.5 FTU. This turbidity was traced to poorly crystallized and extremely fine colloids, which have leached from the surface soils because of the low salinity of the recharge water. Laboratory and field studies showed that gypsum application will reverse the phenomen, but such treatment is uneconomical. This phenomenon is a transient one, and now turbidity outside the recharge area is insignificant from a water quality viewpoint. However, the magnitude of the mass of material in transit through the profile—if stabilized through flocculation or sieving in soil pore space—could greatly change the water transmission and so recharge project performance. However, we have not yet noted this effect at Leaky Acres.

Norberg-King, T.J., E.J. Durhan, G.T. Ankley and E. Robert. "Application of Toxicity Identification Evaluation Procedures to the Ambient Waters of the Colusa Basin Drain, California." *Environmental Toxicology and Chemistry*. V. 10. pp. 891–900. 1991.

Pesticides are applied to the rice fields in the Sacramento Valley to prevent the growth of plants, algae, and insects that reduce rice yields. Following the pesticide application, field water is released into agricultural drains that in turn discharge into the Sacramento River and delta. Rice irrigation is the largest single use of irrigation water in the Sacramento Valley, and because the irrigation water (or rice return) flows are the primary source of drain effluent during the spring and summer (up to 33% of the total flow), these discharges can significantly affect drain water quality and resident aquatic organisms. Acute and chronic toxicity to freshwater organisms (*Ceriodaphnia dubia*) was observed in the drain water during the period that coincides with the initial draining of the fields in 1986, 1987, and 1988. In 1988, a toxicity identification evaluation (TIE) was conducted using *Ceriodaphnia dubia* in an effort to identify the cause of toxicity. Both methyl parathion and carbofuran were identified as possible toxicants. Mixture tests and chronic toxicity tests indicated that the concentrations of methyl parathion and carbofuran in the water sample account for the toxicity observed in *Ceriodaphnia dubia*.

Pahren, H.R. "EPA's Research Program on Health Effects of Wastewater Reuse for Potable Purposes." *Artificial Recharge of Groundwater*. **Butterworth Publishers, Boston. pp. 319–328. 1985.**

One of the many objectives of the Office of Research and Development of the U.S. Environmental Protection Agency (EPA) has been to carry out a relatively small research program on the potential health effects associated with the reuse of renovated wastewater for potable water purposes. This report reviews the research tasks conducted and the results obtained to date.

Research on potable reuse was initiated in 1974 and the federal program funding averaged about $400,000 annually through 1978. Following the 1977 amendments to the Safe Water Drinking Act, which called for special studies on the health implications involved in the reclamation, recycling, and reuse of wastewaters for drinking, funds for reuse research increased. However, the separate program on wastewater reuse was discontinued in 1981. Any activity in the future will be continued as part of the regular drinking water base research program.

Peterson, D.A. "Selenium in the Kendrick Reclamation Project, Wyoming." *Planning Now for Irrigation and Drainage in the 21st Century, Lincoln, Nebraska, 1988.* **(American Society of Civil Engineers, New York, New York, pp. 678–685). 1988.**

Elevated concentrations of selenium in water, bottom sediment, and biota were noted during a reconnaissance investigation of the Kendrick Reclamation Project in central Wyoming. Dissolved-selenium concentrations in 11 of 24 samples of surface or groundwater exceeded the national drinking water standard of 10 µg/L. Bottom sediment samples contained concentrations of several elements, including selenium, that were greater than baseline concentrations in soils of western states. Samples of biota from several trophic levels at four wetlands contained selenium at concentrations associated with physiological problems and abnormalities as reported in laboratory studies and previously published literature.

Petrovic, A.M. "The Fate of Nitrogenous Fertilizers Applied to Turfgrass." *Journal of Environmental Quality.* **V. 19, n. 1. pp. 1–14. 1990.**

Maintaining high-quality surface and groundwater supplies is a national concern. Nitrate is a widespread contaminant of groundwater. Nitrogenous fertilizer applied to turfgrass could pose a threat to groundwater quality. However, a review of the fate of nitrogen applied to turfgrass is lacking, but needed for development of management systems to minimize groundwater contamination. The discussion of the fate of nitrogen applied to turfgrass is developed around plant uptake, atmospheric loss, soil storage, leaching, and runoff. The proportion of the fertilizer nitrogen that is taken up by the turfgrass plant varied from 5 to 74% of applied nitrogen. Uptake was a function of nitrogen release rate and species of grass. Atmospheric loss, by either NH_3 volatilization or denitrification, varied from 0 to 93% of applied nitrogen. Volatilization was generally <36% of applied nitrogen, and can be reduced substantially by irrigation after application. Denitrification was only found to be significant (93% of applied nitrogen) on fine-textured, saturated, warm soils. The amount of fertilizer nitrogen found in the soil plus thatch pool varied as a function of nitrogen source, release rate, age of site, and clipping management. With a soluble nitrogen source, fertilizer nitrogen found in the soil and thatch was 15 to 21% and 21 to 26% of applied nitrogen, respectively, with the higher values reflecting clippings being returned. Leaching losses for fertilizer nitrogen were highly influenced by fertilizer management practices (nitrogen rate, source, and timing), soil texture, and irrigation. Highest leaching losses were reported at 53% of applied nitrogen, but generally were far less than 10%. Runoff of nitrogen applied to turfgrass has been studied to a limited degree, and has been seldom found to occur at concentrations above the federal drinking water standard for NO_3^-. Where turfgrass fertilization poses a threat to groundwater quality, management strategies can allow the turfgrass manager to minimize or eliminate NO_3^- leaching.

Phelps, G.G. *Effects of Surface Runoff and Treated Wastewater Recharge on Quality of Water in the Floridan Aquifer System, Gainesville Area, Alachua County, Florida.* **Geological Survey Water-Resources Investigations Report 87-4099. U.S. Geological Survey, Tallahassee, Florida. 1987.**

Rates of recharge to the Floridan aquifer system at four sites in Alachua County were estimated, and water samples were analyzed to determine if the recharge water had any effects on the water quality of the aquifer. A total of about 33 mgd recharges the upper part of the aquifer system at Haile Sink, Alachua Sink, and drainage wells near Lake Alice. At the Kanapaha Wastewater Treatment Plant, injection wells recharge an average of 6.1 mgd into the lower zone of the system.

The samples of water entering the aquifer system collected at the four sites generally conformed to the drinking water standards recommended by the U.S. Environmental Protection Agency in 1983. Bacteria and nutrient concentrations were more variable in the recharge water than were other constituents. Organic compounds such as diazinon, lindane, and malathion were occasionally detected in all recharge water, but concentrations never exceeded recommended limits.

Bacteria were detected in most wells sampled near the Gainesville recharge sites. The highest counts were from wells near Alachua Sink. At only one site was there a significant difference between the quality of the recharge water and water from the wells sampled, although the recharge water tended to be lower in calcium and iron than water from the Floridan aquifer system. A sample from a well about 150 ft downgradient of a drainage well near Lake Alice consisted of turbid water with a total phosphorus concentration of 75 mg/L and a total nitrogen concentration of 57 mg/L. Water flowing into the drainage well from the lake had a total nitrogen concentration of 1.6 mg/L. Apparently, nutrient-rich suspended sediment in inflow to the drainage well settles out of the water and accumulates in the cavities in the limestone.

Estimated loads entering the aquifer include 3,500 kg/day of chloride, less than 0.43 kg/day lead, 310 kg/day of nitrogen, and 150 kg/day of phosphorus. The effects of the loads were not detected in most monitor wells. Apparently, some of the constituents may settle out, some may be absorbed by the aquifer materials, and the remainder is diluted and dispersed by the extremely large volume of water in the aquifer.

Pierce, R.C. and M.P. Wong. "Pesticides in Agricultural Waters: The Role of Water Quality Guidelines." *Canadian Water Resources Journal.* **V. 13, n. 3. pp. 33–49. 1988.**

Water of good quality is of primary importance to modern agriculture in determining the productivity of crops and the health and marketability of livestock. The quality of water used in agricultural operations can be affected by numerous factors, including pesticide usage. This paper focuses on the relationship between the use of pesticides in Canadian agriculture and the hazards associated with the quality of agricultural waters used in irrigation and livestock watering. The extent and complexity of this problem is assessed initially by examining the overlap between pesticide use and agricultural water use in Canada. The inherent properties of selected pesticides used in Canadian agriculture are highlighted and related to their potential for release to agricultural water supplies. Field and laboratory investigations as related to agricultural water uses are reviewed, and a discussion of pesticide water quality guidelines to ensure protection of agricultural water supplies is provided.

Pitt, W.A.J. Jr. *Effects of Septic Tank Effluent on Ground-Water Quality, Dade County, Florida: An Interim Report.* **Geological Survey Open File Report 74010. U.S. Geological Survey, Tallahassee, Florida. 1974.**

At each of five sites in Dade County, where individual (residence) septic tanks have been in operation for at least 15 years and where septic tank concentration is less than 5 per acre, a drainfield site was selected for investigation to determine the effects of septic tank effluent on the quality of the water in the Biscayne Aquifer.

At each site two sets of multiple-depth wells were drilled. The upgradient wells adjacent to the drainfields in most places were constructed in such a way that the aquifer could be sampled at 10, 20, 30, 40, and 60 ft below land surface. The downgradient wells at each site were 35 ft or more from the upgradient wells in the direction of groundwater flow, and allowed the aquifer to be sampled at various depths.

Except at one site, no fecal coliforms were found below the 10-ft depth. Total coliforms exceeded a count of one colony per mL at the 60-ft depth at two sites. At one site a fecal streptococci count of 53 colonies per mL was found at the 60-ft depth and at another a count of seven colonies was found at the 40-ft depth. The three types of bacteria occur in higher concentrations in the northern areas of the county than in the south. Bacteria concentrations were also higher where the septic tanks were more concentrated.

Pitt, W.A.J. Jr., H.C. Mattraw and H. Klein. *Ground-Water Quality in Selected Areas Serviced by Septic Tanks, Dade County, Florida.* **Geological Survey Open File Report 75-607. U.S. Geological Survey, Tallahassee, Florida. 1975.**

During 1971–74, the U.S. Geological Survey investigated the chemical, physical, bacteriological, and virological characteristics of the groundwater in five selected areas serviced by septic tanks in Dade county, Florida. Periodic water samples were collected from multiple-depth groups of monitor wells ranging in depth from 10 to 60 ft at each of the five areas. Analyses of groundwater from baseline water quality wells in inland areas remote from urban development indicated that the groundwater is naturally high in organic nitrogen, ammonia, organic carbon, and chemical oxygen demand. Some enrichment of groundwater with sodium provided a possible key to differentiating septic tank effluent from other urban groundwater contaminant sources. High ammonia nitrogen, phosphorus, and the repetitive detection of fecal coliform bacteria were characteristic of two 10-ft monitor wells that consistently indicated the presence of septic tank effluent in groundwater. Dispersion, dilution, and various chemical processes have presumably prevented accumulation of septic tank effluent at depths greater than 20 ft, as indicated by the 65 types of water analyses used in the investigation. Fecal coliform bacteria were present on one or two occasions in many monitor wells, but the highest concentration—1,600 colonies/100 mL—was related to stormwater infiltration rather than septic tank discharge.

Areal variations in the composition and the hydraulic conductivity of the sand and limestone aquifer had the most noticeable influence on the overall groundwater quality. The groundwater in the more permeable limestone in south Dade County near Homestead contained low concentrations of septic tank–related constituents, but higher concentrations of dissolved sulfate and nitrate. The groundwater in north Dade County, where the aquifer is less permeable, contained the highest dissolved iron, manganese, COD, and organic carbon.

Power, J.F. and J.S. Schepers. "Nitrate Contamination of Groundwater in North America." *Agriculture, Ecosystems and Environment.* **V. 26, n. 3–4. pp. 165–187. 1989.**

Groundwater serves as the primary domestic water supply for over 90% of the rural population and 50% of the total population of North America. Consequently, protection of groundwater from contamination is of major concern. This paper reviews the problem of controlling nitrate pollution of groundwater in North America. Nitrates in groundwater originate from a number of nonpoint sources, including geological origins, septic tanks, improper use of animal manures, cultivation (especially fallowing), precipitation, and fertilizers. Accumulation of nitrate nitrogen in groundwater is probably attributed to different regions. Major areas of nitrate pollution often occur under irrigation because leaching is required to control salt accumulation in the root zone. In the last few decades, areas under irrigation and the use of nitrogen fertilizers have increased greatly, and both of these have probably contributed to groundwater nitrate problems. Use of known best management practices (irrigation scheduling; fertilization based on calibrated soil tests; conservation tillage; acceptable cropping practices; recommended manuring rates) has been demonstrated to be highly effective in controlling leaching of nitrates. Government policies are needed that will encourage and reward the use of the best management practices that help control nitrate accumulations in groundwater.

Pruitt, J.B., D.E. Troutman and G.A. Irwin. *Reconnaissance of Selected Organic Contaminants in Effluent and Ground Water at Fifteen Municipal Wastewater Treatment Plants in Florida.* **Geological Survey Water-Resources Investigations Report 85-4167. U.S. Geological Survey, Tallahassee, Florida. 1985.**

Results of a 1983–84 reconnaissance of 15 municipal wastewater treatment plants in Florida indicated that effluent from most of the plants contains trace concentrations of volatile organic compounds. Chloroform was detected in the effluent at 11 of the 15 plants and its common occurrence was likely the result of chlorination. The maximum concentration of chloroform detected in the effluent sampled was 120 µg/L. Detectable concentrations of selected organophosphorus insecticides were also common. For example, diazinon was detected in the effluent at 12 of the 15 plants, with a maximum concentration of 1.5 µg/L. Organochlorine insecticides, primarily lindane, were detected in the effluent at 8 of the 15 plants, with a maximum concentration of 1.0 µg/L.

Volatile compounds, primarily chloroform, were detected in water from monitor wells at four plants, and organophosphorus insecticides, primarily diazinon, were present in the groundwater at three treatment plants. Organochlorine insecticides were not detected in any samples from monitor wells. Based on the limited data available, this cursory reconnaissance suggests that the organic contaminants commonly occurring in the effluent of many of the treatment plants are not transported into the local groundwater.

Ragone, S.E. *Geochemical Effects of Recharging the Magothy Aquifer, Bay Park, New York, with Tertiary-Treated Sewage.* **Geological Survey Professional Paper 751-D. U.S. Government Printing Office, Washington, D.C. 1977.**

A groundwater deficit of 93.5 to 123 mgd (4.10 to 5.39 m³/s) has been predicted for Nassau County, N.Y., by the year 2000 in a state report. Because of the predicted deficit, the U.S. Geological Survey, in cooperation with the Nassau County Department of Public Works, began an experimental deep-well recharge program in 1968. Thirteen recharge tests using tertiary-treated sewage (reclaimed water) and six tests using water from the domestic supply (city water) were completed between 1968 and 1973. Recharge was through an 18-in. (46-cm) diameter recharge well screened in the Magothy aquifer between depths of 418 and 480 ft (127 and 146 m) below land surface. Recharge rates ranged from about 200 to 400 gpm (13 to 25 L/s). In the longest test, reclaimed water was injected during 84.5 days of a 199-day period.

Although the iron concentration of native water in the recharge zone and reclaimed water is less than 0.5 mg/L, the iron concentration of samples collected from observation wells 20, 100, and 200 ft (6.1, 30, and 61 m) from the recharge well, and screened in the zone of recharge, approached 3 mg/L at times. Iron mass-balance calculations indicate that dissolution of pyrite and marcasite (FeS_2) in the aquifer are the only known sources of iron that could explain the observed increase. Within a 20-ft (6.1-m) radius of the recharge well, dissolved oxygen in the reclaimed water oxidizes pyrite and releases Fe^{+2} (ferrous iron) to solution. However, the amount of iron in water continues to increase with distance from the recharge well even though dissolved oxygen is no longer present in water reaching the 20-ft (6.1 m) radius; the mechanism by which iron continues to be dissolved is not quantitatively understood.

Some cation exchange also occurs during recharge. Loss of ammonium and potassium cations in the water was balanced by an increase in H^+, which at times caused pH to decrease by more than 1 pH unit.

Tertiary treatment removes 90 to 98% of the phosphate, MBAS (methylene blue active substances), and COD (chemical oxygen demand), leaving an average of 0.17, 0.07, and 9 mg/L, respectively. During recharge, phosphate concentrations remain at native-water levels at the 20-, 100-, and 200-ft (6.1-, 30-, and 61-m) observation wells, which indicates phosphate retention by the aquifer. Some MBAS and COD are retained at the 100- and 200-ft (30- and 61-m) wells, presumably by adsorption reactions.

Ragone, S.E., H.F.H. Ku and J. Vecchioli. "Mobilization of Iron in Water in the Magothy Aquifer During Long-Term Recharge with Tertiary-Treated Sewage, Bay Park, New York." *Journal Research U.S. Geological Survey.* **V. 3, n. 1. pp. 93–98. 1975.**

Tertiary-treated sewage (reclaimed water) has been recharged by well into the Magothy aquifer at Bay Park, NY, intermittently since 1968. The longest of 13 recharge tests, the subject of this report, lasted 84.5 days. This was sufficient time for the reclaimed water to reach an observation well 200 ft (61 m) from the recharge well. Although the iron concentrations of the reclaimed water and the native water were less than 0.4 mg/L, the iron concentrations of samples from observation wells 20,000 and 200 ft (6,300, and 61 m) from the recharge well at times approached 3 mg/L. Source of the iron is pyrite that is native to the aquifer.

Ragone, S.E. and J. Vecchioli. "Chemical Interaction during Deep Well Recharge, Bay Park, New York." *Ground Water*. **V. 13, n. 1. Reprint. 1975.**

The U.S. Geological Survey, in cooperation with the Nassau County Department of Public Works, recharged tertiary-treated sewage (reclaimed water) into the Magothy aquifer in 13 recharge experiments between 1968 and 1973. The recharge resulted in a degradation in water quality with respect to iron concentration and pH. Iron concentration increased from the range 0.14 to 0.30 mg/L to as much as 3 mg/L at the 20-, 100-, and 200-ft or 6.1-, 30-, and 61-m observation wells as the reclaimed water displaced native water. The increase was presumably a result of pyrite dissolution. The pH of the water decreased from the range 5.22 to 5.72 to a low of about 4.50, predominantly as a result of cation-exchange reactions.

Ramsey, R.H. III, J. Borreli and C.B. Fedler. "The Lubbock, Texas, Land Treatment System." *Irrigation Systems for the 21st Century, Portland, Oregon, 1987.* **(American Society of Civil Engineers, New York, New York, pp. 352– 361). 1987.**

The land treatment system at Lubbock, Texas provides an excellent model for studying the response of a system to growth. It also provides insights and justification for current criteria used to design slow-rate land treatment systems. During the 62 years of operation, the Lubbock Land Treatment System responded to a substantial increase in the volume of the effluent and changes in environmental concern for ground-water pollution. While certain problems persist, they have demonstrated that short-comings in the system design can be turned into positive assets. Groundwater from beneath the treatment farms is used to provide flat water recreation and irrigation of city parks and cemeteries.

Razack, M., C. Drogue and M. Baitelem. "Impact of an Urban Area on the Hydro-chemistry of a Shallow Groundwater (Alluvial Reservoir) Town of Narbonne, France." *Hydrological Processes and Water Management in Urban Areas, Duisburg, Federal Republic of Germany, 1988.* **(International Hydrological Programme, UNESCO, pp. 487–494). 1988.**

The Roman-founded urban area of Narbonne in Southern France is built upon a shallow groundwater. This reservoir, whose thickness ranges from 10 to 30 m, is composed of various packed materials and quaternary deposits. A hydrochemical survey carried out during the summer of 1984 and the winter of 1985 showed an important impact of the urban activity on the groundwater quality. This impact is expressed through the superimposition beneath the city of a chemistry of exogenous elements coming from urban activity (SO_4, NO_3) and of a natural chemistry (Na, Cl), both displaying different aerial patterns.

Rea, A.H. and J.D. Istok. "Groundwater Vulnerability to Contamination: A Literature Review." *Irrigation Systems for the 21st Century, Portland, Oregon, 1987.* **(American Society of Civil Engineers, New York, New York, pp. 362–367). 1987.**

Methods are needed to allow regulatory agencies and resource managers to predict, from readily available data, the potential for groundwater contamination problems. Regional maps developed from these methods can aid in planning and allocation of resources. The literature was searched for methods capable of filling this need. Six methods were selected and compared using hypothetical hydrogeologic settings. Based on the results obtained, which varied considerably between methods, none of the methods are completely satisfactory. Of the reviewed methods, "DRASTIC" seemed to be most suitable.

Reichenbaugh, R.C. *Effects on Ground-Water Quality from Irrigating Pasture with Sewage Effluent Near Lakeland, Florida.* **Geological Survey Water-Resources Investigations 76-108. U.S. Geological Survey, Tallahassee, Florida. 1977.**

Since 1969, on the average, 25,000 gal (94,600 L/day) of domestic secondary-treated effluent has been used each day to supplement irrigation of 30 ac (12 ha) of grazed pasture north of Lakeland, in west-central Florida. The U.S. Geological Survey began a study of the site several months after sprinkler application of the effluent to the Myakka sands (well sorted, fine, acid) was started. The site, on the south shore of Lake Gibson, is underlain by as much as 60 ft (18 m) of sand, sandy clay, and clay containing the water-table aquifer, and two relatively unimportant confined aquifers, which in turn are underlain by the confined Floridan aquifer.

Monitor wells were constructed to various depths in clusters near the effluent-irrigated pasture. The water table in the surficial aquifer varied from 1 to 3.3 ft (0.3 to 1.0 m) below the land surface. Groundwater quality was evaluated by analysis of water samples collected three times over a 1-year period.

Groundwater beneath the irrigated pasture showed slight increases in cations and anions which are attributed to irrigation with the effluent. The concentration of total nitrogen (predominantly ammonia and organic nitrogen) was reduced to less than 20% of that in the upper 8 ft (2.4 m) of pasture soils, and there was no increase in concentration below 20 ft (6.1 m), or in downgradient groundwater. There was no evidence of phosphorus or carbon contamination of groundwater at the site. Though small numbers of bacteria were noted in some samples from nine wells, most were of the coliform group. Only four wells yielded samples containing bacteria of probable fecal origin—one colony per 100 mL in each sample.

There was no detected accumulation of solids at the soil surface. Organic carbon, pH, and Kjeldahl nitrogen concentrations of the soil in the irrigated pasture were only slightly higher when compared to soil outside the pasture. As of 1972, the low-rate application of the effluent to the pasture apparently has had little effect on the soil and groundwater.

Reichenbaugh, R.C., D.P. Brown and C.L. Goetz. *Results of Testing Landspreading of Treated Municipal Wastewater at St. Petersburg, Florida.* **Geological Survey Water-Resources Investigations Report 78-110. U.S. Geological Survey, Tallahassee, Florida. 1979.**

Chlorinated secondary-treated effluent was used to irrigate a grassed 4-ac site at rates of 2 and 4 in./week for periods of 11 and 14 weeks, respectively. Part of the site was drained by tile lines 5 ft below land surface. Chemical and bacteriological changes in the acidic groundwater in the shallow aquifer and in the effluent from the drains were studied.

Irrigation of the drained plot resulted in rapid passage of the applied wastewater through the soil and, consequently, poor nitrogen removal. The rapid percolation permitted nitrification but prevented denitrification. Thus, the effluent from the drains contained as much as 5.2 mg/L nitrate-nitrogen. Irrigation of the undrained plot resulted in more extensive nitrogen removal.

Total phosphorus in the shallow groundwater at the site increased from a maximum of 1.4 mg/L before irrigation to as much as 5 mg/L in the groundwater 5 ft below land surface.

Concentrations of nitrogen and phosphorus did not increase in groundwater downgradient from the site, although increased chloride concentrations demonstrated downgradient migration of the applied wastewater.

Prior to irrigation, total coliform bacteria were not detected in groundwater at the site. After irrigation, total and fecal coliforms were detected in the groundwater at the site and downgradient. The nitrifying bacteria *Nitrosomonas* and *Nitrobacter* at the irrigated site were most abundant at the soil surface; their numbers decreased with depth.

Rein, D.A., G.M. Jamesson and R.A. Monteith. "Toxicity Effects of Alternative Disinfection Processes." *Water Environment Federation 65th Annual Conference & Exposition, New Orleans, Louisiana, 1992.* **(Water Environment Federation, Alexandria, Virginia, pp. 461–470). 1992.**

Chlorination/dechlorination, ultraviolet irradiation, and ozonation were evaluated in side-by-side pilot/bench-scale tests at the Akron, Ohio Water Pollution Control Station. The objective of the evaluation was to investigate the effect of these disinfection processes on final effluent toxicity.

Six sets of chronic and two sets of acute whole effluent toxicity tests were conducted using *P. promelas* and *C. dubia*. After the first two sets of tests, it became apparent that differences between the processes could only be seen in 100% effluent samples. The test protocol was then modified to compare the relative toxic response in terms of survival and reproduction of *C. dubia* in 100% effluent samples.

Although very little statistically estimated toxicity was found in any of the process effluents, the chlorination/dechlorination process effluent consistently produced a greater relative toxic response than either ultraviolet irradiation or ozonation. The chlorination/dechlorination process produced the greatest relative toxic response for offspring produced and/or survival of *C. dubia* in seven of eight tests sets using 100% effluent samples.

Rice, R.C., D.B. Jaynes and R.S. Bowman. "Preferential Flow of Solutes and Herbicide under Irrigated Fields." *Transactions of the American Society of Agricultural Engineers.* **V. 34, n. 2. pp. 914–918. 1991.**

Over the past several years, there has been an increasing concern of groundwater contamination from agricultural chemicals. Until recently it was generally believed that pesticides would not move to the groundwater. Starting in the mid-70s more cases of pesticide contamination were reported. This article discusses recent experiments where accelerated leaching of solutes and an herbicide were observed under intermittent flood and sprinkler irrigation. Preferential flow phenomena resulted in solute and herbicide velocities of 1.6 to 2.5 times faster than that calculated by traditional water balance methods and piston flow model. Little preferential flow was observed under continuously flooded conditions on a loam soil. Generally, preferential flow is thought to occur in coarse-grained soils, cracked soils, or in macropores such as root or worm holes. The bypass that we observed was in sandy loam and sandy soils with little or no structure. Understanding the preferential flow phenomenon is necessary when predictive flow models are used.

Ritter, W.F., F.J. Humenik and R.W. Skaggs. "Irrigated Agriculture and Water Quality in the East." *Journal of Irrigation and Drainage Engineering.* **V. 115, n. 5. pp. 807–821. 1989.**

The northeastern and Appalachian states have a diverse array of geology, soils, and climate. Irrigation is concentrated in a few states, with the largest irrigation area in the Coastal Plain soils. Most of these soils are sandy and very susceptible to leaching. The groundwater recharge area in the Coastal Plain is directly above the aquifer. Most of the increase in irrigation has been to irrigate corn in Delaware, Maryland, and Virginia. Groundwater studies have been conducted in Delaware, Maryland, and New York in irrigated regions. Nitrate and aldicarb leaching has occurred on Long Island, New York, where potatoes are grown. Poultry manure is the largest source of nitrate contamination of the water table aquifer on the Delmarva Peninsula in Maryland. Both pesticide and nitrate leaching under irrigation have been studied in Delaware. A total water management system that can be used for both drainage and subsurface irrigation has been developed in North Carolina. The system will increase crop yields, and has the potential for reducing nitrates by water table control.

Ritter, W.F., R.W. Scarborough and E.M. Chirnside. "Nitrate Leaching Under Irrigation on Coastal Plain Soil." *Journal of Irrigation and Drainage Engineering.* **V. 117, n. 4. pp. 490–502. 1991.**

The effect of irrigation and nitrogen management on groundwater quality was evaluated for four years on a Sassafras sandy loam Coastal Plain soil. Applying the greatest portion of the nitrogen by side-dressing and by fertigation were compared. Maintaining optimal soil moisture and partial irrigation (applying one-half the water as optimal irrigation) were the water management practices investigated. Nitrate concentrations increased in the groundwater for all nitrogen and irrigation-management practices. The mass of the nitrate leached was related to the drainage volume. In all but one year the largest mass of nitrate was leached during the fall and winter months, when the largest amount of recharge occurs. Very little nitrate was leached during the growing season for any nitrogen or irrigation management practice except for one year when 30 cm of rainfall occurred in August. The mass of nitrate leached during that growing season ranged from 33.9 kg/ha for partial irrigation to 139.0 kg/ha for full irrigation.

Robinson, J.H. and H.S. Snyder. "Golf Course Development Concerns in Coastal Zone Management." *Coastal Zone '91: Proceedings of the Seventh Symposium on Coastal and Ocean Management, Long Beach, California, 1991.* **(American Society of Civil Engineers, New York, New York, pp. 431–442). 1991.**

The rapid growth of golf course development in South Carolina's coastal zone presents new challenges in protecting coastal water and wetlands. While filling or dredging of wetland resources for golf course development is a practice of the past, golf course designers make every effort to minimize their proximity to wetland resource areas. Coastal zone management concerns associated with golf course development include the protection of adjacent wetland resources from (1) nutrient- and chemical-laden stormwater runoff, (2) aerosol from "fertigation,"—a mixture of fertilizer and irrigation water—and treated effluent irrigation systems, and (3) physical impacts associated with wetland crossings, play-through areas, and player intrusion. This paper first provides a brief overview of current literature associated with the use of chemicals on golf courses and their impacts on man and the coastal environment. This paper then focuses on best management practices which can be utilized in the physical design and management of golf courses to minimize impacts on coastal resources, drawing upon examples from golf courses recently constructed or currently under development in coastal South Carolina.

Rosenshein, J.S. and J.J. Hickey. "Storage of Treated Sewage Effluent and Storm Water in a Saline Aquifer, Pinellas Peninsula, Florida." *Ground Water.* **V. 15, n. 4. pp. 284–293. 1977.**

The Pinellas Peninsula, an area of 750 square kilometers (290 square miles) in coastal west-central Florida, is a small hydrogeologic replica of Florida. Most of the peninsula's water supply is imported from well fields as much as 65 km (40 mi) inland. Stresses on the hydrologic environment of the peninsula and on adjacent water bodies, resulting from intensive water-resources development and waste discharge, have resulted in marked interest in subsurface storage of wastewater (treated effluent and untreated stormwater) and in future retrieval of stored water for nonpotable use. If subsurface storage is approved by regulatory agencies, as much as 265 ML/day (70 mgd) of wastewater could be stored underground within a few years, and more than 565 ML/day (150 million mgd) could be stored in about 25 years. This storage would constitute a large resource of nearly fresh water in the saline aquifers underlying about 250 km^2 (200 mi^2) of the Peninsula.

The upper 1,060 m (3,480 ft) of the rock column underlying four test sites on the Pinellas Peninsula have been explored. The rocks consist chiefly of limestone and dolomite. Three moderately to highly transmissive zones, separated by leaky confining beds (low-permeability limestone) from about 225 to 380 m (740 to 1,250 ft) below mean sea level, have been identified in the lower part of the Floridan aquifer in the Avon Park Limestone. Results of withdrawal and injection tests in Pinellas County indicate that the middle transmissive zone has the highest estimated transmissivity—about 10 times other reported values. The chloride concentration of water in this zone, as well as in the Avon Park Limestone in Pinellas Peninsula, is about 19,000 mg/L. If subsurface storage is approved and implemented, this middle zone probably would be used for storage of the wastewater and the site would become the most extensively used in Florida for this purpose.

Sabatini, D.A. and T.A. Austin. "Adsorption, Desorption and Transport of Pesticides in Groundwater: A Critical Review." *Planning Now for Irrigation and Drainage in the 21st Century, Lincoln, Nebraska, 1988.* **(American Society of Civil Engineers, New York, New York, pp. 571–579). 1988.**

Adsorption and desorption are major mechanisms affecting the transport and fate of pesticides in groundwater. Equilibrium, chemical nonequilibrium and physical nonequilibrium adsorption relationships for predicting pesticide transport are reviewed. Hysteresis of desorption and relationships for predicting linear equilibrium coefficients are discussed.

Sabol, G.V., H. Bouwer and P.J. Wierenga. "Irrigation Effects in Arizona and New Mexico." *Journal of Irrigation and Drainage Engineering.* **V. 113, n. 1. pp. 30–57. 1987.**

Irrigated agriculture accounts for about 90% of all water consumption in both Arizona and New Mexico. More than 50% of this water is pumped from groundwater sources. Some portion of the applied irrigation water is returned to the groundwater supply through deep percolation. Several field studies have been conducted in these states to measure the quantity and quality of water that is recharging the aquifer. These studies indicate that groundwater quality in Arizona and New Mexico has been deleteriously affected in deep supplies. The magnitude and time rate of groundwater quality changes are a function of irrigation management practice, fertilizer and pesticide applications, quality of irrigation water, rate of groundwater level decline, presence of perched zones that intercept percolating water, proximity to surface water supplies, leakage through and along well casings, and soil salinity.

Salo, J.E., D. Harrison and E.M. Archibald. "Removing Contaminants by Groundwater Recharge Basins." *Journal of the American Water Works Association.* **V. 78, n. 9. pp. 76–81. 1986.**

The effects of urban runoff used to recharge groundwater basins in the Fresno, Calif., area are discussed. The study, which was part of the U.S. Environmental Protection

Agency's Nationwide Urban Runoff Program, was undertaken to determine the environmental effects on groundwater quality and to identify management practices that would mitigate adverse effects. No deterioration of groundwater quality because of recharge with runoff was found, although the authors recommend a more thorough investigation of effects of recharge with runoff from industrial sites.

Schiffer, D.M. *Effects of Three Highway-Runoff Detention Methods on Water Quality of the Surficial Aquifer System in Central Florida.* **Geological Survey Water-Resources Investigations Report 88-4170. U.S. Geological Survey, Tallahassee, Florida. 1989.**

Water quality of the surficial aquifer system was evaluated at one exfiltration pipe, two ponds (detention and retention), and two swales in central Florida, representing three runoff detention methods, to detect any effects from infiltrating highway runoff. Concentrations of major ions, metals, and nutrients were measured in groundwater and bottom sediments from 1984 through 1986.

At each study area, concentrations in groundwater near the structure were compared to concentrations in groundwater from an upgradient control site. Groundwater quality data also were pooled by detention method and statistically compared to detect any significant differences between methods.

Analysis of variance of the rank-converted water quality data at the exfiltration pipe indicated that mean concentrations of 14 to 26 water quality variables are significantly different among sampling locations (the pipe, unsaturated zone, saturated zone, and the control well). Most of these differences are between the unsaturated zone and other locations. Only phosphorus is significantly higher in groundwater near the pipe than in groundwater at the control well.

Analysis of variance of rank-converted water quality data at the retention pond indicated significant differences in 14 to 25 water quality variables among sampling locations (surficial aquifer system, intermediate aquifer, pond, and the control well), but mean concentrations in groundwater below the pond were never significantly higher than in groundwater from the control well. Analysis of variance results at other study areas indicated few significant differences in water quality among sampling locations.

Values of water quality variables measured in groundwater at all study areas generally were within drinking water standards. The few exceptions included pH (frequently lower than the limit of 6.5 at one pond and both swales), and iron, which frequently exceeded 300 µg/L in groundwater at one swale and the detention pond.

Large concentrations of polyaromatic hydrocarbons were measured in sediments at the retention pond, but qualitative analysis of organic compounds in groundwater from three wells indicated concentrations of only 1 to 5 µg/L at one site, and below detection level (1 µg/L) at the other two sites. This may be an indication of immobilization of organic compounds in sediments.

Significant differences for most variables were indicated among groundwater quality data pooled by detention method. Nitrate nitrogen and phosphorus concentrations were highest in groundwater near swales and exfiltration pipe; the Kjeldahl nitrogen was highest near ponds. Chromium, copper, and lead concentrations in groundwater were frequently below detection levels at all study areas, and no significant differences among detention methods were detected for metal concentrations, with the exception of iron. High iron concentrations in groundwater near the detention pond and one swale most likely were naturally occurring and unrelated to highway runoff.

Results of the study indicate that natural processes occurring in soils attenuate inorganic constituents in runoff prior to reaching the receiving groundwater. However, organic compounds detected in sediments at the retention pond indicate a potential problem that may eventually affect the quality of the receiving groundwater.

Schmidt, K.D. and I. Sherman. "Effect of Irrigation on Groundwater Quality in California." *Journal of Irrigation and Drainage Engineering.* **V. 113, n. 1. pp. 16–29. 1987.**

Deep percolation of irrigation return flow is a major source of recharge beneath most irrigated areas in California. Tile drainage, soils, water in the vadose zone, and shallow groundwater have been studied. Nitrate, salinity, and several pesticides have received the most attention. Numerous parts of the San Joaquin Valley have been investigated, as well as parts of the Sacramento Valley, Imperial Valley, Los Angeles Basin, and several other valleys. The results of the studies indicate that irrigation return flow usually exerts a substantial impact on groundwater quality. High nitrate contents in groundwater beneath irrigated areas are often a result of irrigation. In addition, extensive pollution of shallow groundwater in parts of the San Joaquin Valley have been caused by use of the pesticide DBCP.

Schneider, B.J., H.F.H. Ku and E.T. Oaksford. *Hydrologic Effects of Artificial-Recharge Experiments with Reclaimed Water at East Meadow, Long Island, New York.* **Geological Survey Water Resources Investigations Report 85-4323. U.S. Geological Survey, Denver, Colorado. 1987.**

Artificial-recharge experiments were conducted at East meadow from October 1982 through January 1984 to evaluate the degree of groundwater mounding and the chemical effects of artificially replenishing the groundwater system with tertiary-treated wastewater. More than 800 million gallons of treated effluent was returned to the upper glacial aquifer through recharge basins and injection wells in the 15-month period.

Reclaimed water was provided by the Cedar Creek advanced wastewater treatment facility in Wantagh, 6 miles away. The chlorinated effluent was pumped to the recharge facility, where it was fed to basins by gravity flow and to injection wells by pumps. An observation well network was installed at the recharge facility to monitor both physical and chemical effects of reclaimed water on the groundwater system.

Observations during the recharge tests indicate that the two most significant factors in limiting the rate of infiltration through the basin floor were the recharge test duration and the quality of the reclaimed water. Head buildup in the aquifer beneath the basins ranged from 4.3 to 6.7 ft, depending on the quantity and duration of water application. Head buildup near the injection wells within the aquifer ranged from 0.3 to 1.2 ft. The head buildup in the injection wells is attributed to biological, physical, and chemical actions, which can operate separately or together. Recharge basins provided a more effective means of moving large quantities of reclaimed water into the aquifer than injection wells.

Two basins equipped with central observation manholes permit the acquisition of data on the physical and chemical processes that occur within the unsaturated zone during recharge. Results of 3-day and 176-day ponding tests in basins 3 and 2, respectively, indicate that reclaimed water is relatively unchanged chemically by percolation through the unsaturated zone because (1) the sand and gravel of the upper glacial aquifer is unreactive, (2) the water moves to the water table rapidly, and (3) the water is highly treated before recharge.

The quality of water in the aquifer zones affected by recharge improved, on the whole. Groundwater concentrations of nitrate-nitrogen and several low-molecular-weight hydrocarbons, although significantly above drinking water standards before recharge, decrease to well within drinking water standards as a direct result of recharge. Sodium and chloride concentrations increased above background levels as a result of recharge, but remained well within drinking water standards and the New York State effluent standards established for this groundwater recharge study.

Seaburn, G.E. and D.A. Aronson. *Influence of Recharge Basins on the Hydrology of Nassau and Suffolk Counties, Long Island, New York.* **Geological Survey Water Supply Paper 2031. U.S. Government Printing Office, Washington, D.C. 1974.**

An investigation of recharge basins on Long Island was made by the U.S. Geological Survey in cooperation with the New York State Department of Environmental Conservation, the Nassau County Department of Public Works, and the Suffolk County Water Authority. The major objectives of the study were to (1) catalog basic physical data on the recharge basins in use on Long Island, (2) measure quality and quantity of precipitation and inflow, (3) measure infiltration rates at selected recharge basins, and (4) evaluate regional effects of recharge basins on the hydrologic system of Long Island. The area of study consists of Nassau and Suffolk Counties—about 1,370 mi^2—in eastern Long Island, N.Y.

Recharge basins, numbering more than 2,100 on Long Island in 1969, are open pits on moderately to highly permeable sand and gravel deposits. These pits are used to dispose of storm runoff from residential, industrial, and commercial areas, and from highways, by infiltration of the water through the bottom and sides of the basins.

The hydrology of three recharge basins on Long Island—Westbury, Syosset, and Deer Park basins—was studied. The precipitation-inflow relation showed that the average percentages of precipitation flowing into each basin were roughly equivalent to the average percentages of impervious areas in the total drainage areas of the basins. Average percentages of precipitation flowing into the basins as direct runoff were 12% at the Westbury basin, 10% at the Syosset basin, and 7% at the Deer Park basin. Numerous open-bottomed stormwater catchbasins at Syosset and Deer Park reduced the proportion of inflow to those basins, as compared with the Westbury basin, which has only a few open-bottomed catchbasins.

Inflow hydrographs for each basin typify the usual urban runoff hydrograph—steeply rising and falling limbs, sharp peaks, and short time bases. Unit hydrographs for the Westbury and the Syosset basins are not expected to change; however, the unit hydrograph for the Deer Park basin is expected to broaden somewhat as a result of additional future house construction within the drainage area.

Infiltration rates averaged 0.9 fph (feet per hour) for 63 storms between July 1967 and May 1970 at the Westbury recharge basin, 0.8 fph for 22 storms from July 1969 to September 1970 at the Syosset recharge basin, and 0.2 fph for 24 storms from March to September 1970 at the Deer Park recharge basin. Low infiltration rates at Deer Park resulted mainly from (1) a high percentage of eroded silt, clay, and organic debris washed in from construction sites in the drainage area, which partly filled the interstices of the natural deposits, and (2) a lack of well developed plant-root system on the younger basin, which would have kept the soil zone more permeable.

The apparent rate of movement of stormwater through the unsaturated zone below the unsaturated zone below each basin averaged 5.5 fph at Westbury, 3.7 fph at Syosset, and 3.1 fph at Deer Park. The rates of movement for storms during the warm months (April through October) were slightly higher than average, probably because the recharging water was warmer than it was during the rest of the year, and therefore was slightly less viscous.

On the average, a 1-in. rainfall resulted in a peak rise of the water table directly below each basin of 0.5 ft; a 2-in. rainfall resulted in a peak rise of about 2 ft. The mound commonly dissipated within 1 to 4 days at Westbury, 7 days to more than 15 days at Syosset, and 1 to 3 days at Deer Park, depending on the magnitude of the peak buildup.

Average annual groundwater recharge was estimated to be 6.4 ac-ft at the Westbury recharge basin, 10.3 ac-ft at the Syosset recharge basin, and 29.6 ac-ft at the Deer Park recharge basin.

Chemical composition of precipitation at Westbury, Syosset, and Deer Park drainage areas was similar: hardness of water ranged from 6 to 56 mg/L (as calcium and magnesium hardness), dissolved solids content ranged from 21 to 124 mg/L, and pH ranged from 5.9 to 6.6. Calcium was the predominant cation, and sulfate and bicarbonate were the predominant anions. Atmospheric dust and gaseous sulfur compounds associated with the Northeast's urban environment mainly account for this combination of ions in precipitation.

Chemical composition of the inflow to the basins was also similar in each of the three basins. In general, hardness of the water samples collected at Westbury, Syosset, and Deer Park recharge basins in 1970 was less than 50 mg/L (as calcium and magnesium hardness), and dissolved solids content was less than 100 mg/L. The pH ranged from 6.1 to 7.4. The concentrations of most constituents in inflow were greater than those in precipitation; precipitation contributed 70 to 88% of the loads of dissolved constituents in the inflow.

Only 3 of 11 pesticides sought by chemical analysis were detected. A maximum DDT concentration of 0.08 μg/L was determined for an inflow sample to Westbury recharge basin. Concentrations of other pesticides were 0.02 μg/L or less.

Total concentration of pesticides detected in the soil layers on the floors of each basin generally ranged from 0.4 to 40 mg/L. The greater organic content of the soil layers, compared with that of the underlying natural deposits, suggests that pesticides as well as other organic material are effectively reduced or removed from the infiltrating water in the soil layer.

Groundwater recharge from precipitation through the total area (73,000 acres) drained by 2,124 recharge basins in operation in 1969 was estimated to be 166,000 ac-ft per year, or about 148 million gpd. Groundwater recharge in the areas where recharge basins are used is probably equivalent to or may slightly exceed recharge under natural conditions.

Shirmohammadi, A. and W.G. Knisel. "Irrigated Agriculture and Water Quality in the South." *Journal of Irrigation and Drainage Engineering*. V. 115, n. 5. pp. 791–806. 1989.

Irrigated agriculture in the humid region has resulted in more intensive management, including crop production and an associated increase in fertilizer and pesticide use. Multiple cropping in most of the southeast (Alabama, Florida, Georgia, and South Carolina) and Delta (Arkansas, Louisiana, and Mississippi) states increases the demand for water and agricultural chemicals. Pesticide usage in the 48 contiguous states and the District of Columbia totaled 299,892,159 kg of active ingredient (AI) by 1982. Agricultural chemicals may percolate to aquifers in some soils and geologic formations, resulting in groundwater contamination. Groundwater fluctuations are related to irrigation. Groundwater quality data are used to show the trend in quality related to irrigated agriculture and cropping systems. Areas with specific groundwater problems, such as saltwater intrusion and pesticide levels, are identified. A total of 17 pesticides have been reported in groundwater in the United States, and four of these were found in the southeast and Delta states. Data show that less than 1% of wells sampled in the southeast and Delta states had nitrate concentrations exceeding 10 mg/L (drinking water standard). Degradation of surface water quality relative to irrigation is discussed.

Smith, S.O. and D.H. Myott. "Effects of Cesspool Discharge on Ground-Water Quality on Long Island, N.Y." *Journal of the American Water Works Association*. V. 67, n. 8. pp. 456–458. 1975.

Large amounts of household wastes, discharged through cesspools, have resulted in deterioration of groundwater quality on Long Island. Although nitrate pollution poses the greatest threat to the Island's water supplies, other constituents derived from cesspool leachings are increasing. Municipal sewering projects that have been undertaken as a solution are discussed.

Spalding, R.F. and L.A. Kitchen. "Nitrate in the Intermediate Vadose Zone Beneath Irrigated Cropland." *Ground Water Monitoring Review*. V. 8, n. 2. pp. 89–95. 1988.

More than 1000 ft of fine-textured, unsaturated zone core beneath nitrogen-fertilized and irrigated farmland was collected, leached, and analyzed for nitrate-nitrogen. Fertility plots treated with 200, 300 and 400 lb N/ac/yr accumulated significant quantities of nitrate-nitrogen in the vadose zone below the crop rooting zone. The average nitrate-nitrogen concentration approximately doubled with each 100 lb N/ac/yr increment above the 100 lb N/ac/yr treatment. Nitrate loading estimates for the plots treated with 400 lb N/ac/yr indicate that over 1200 lb N/ac was in the vadose zone beneath the crop rooting zone. In 15 years, the nitrate moved vertically at least 60 ft through these fine-textured, unsaturated sediments. As much as 600 lb N/ac have accumulated in the vadose zone under independent corn producers' fields.

Vadose zone sampling is effective in predicting future nonpoint nitrate-contaminated areas.

Squires, R.C., G.R. Groves and W.R. Johnston. "Economics of Selenium Removal from Drainage Water." *Journal of Irrigation and Drainage Engineering.* **V. 115, n. 1. pp. 48–57. 1989.**

A treatment system consisting of biological reactors and microfiltration has been developed to remove soluble selenium species from agricultural drainage water. The process was evaluated over a two-year period, and the reactor configurations and specific removal rates of nitrate and selenium were optimized. Trials on the operation of a pilot solar salt works to concentrate the detoxified water after treatment to recover salts were also carried out. The treatment process reduced the selenium concentration of the drainage water from over 500 mg/L to 10–50 mg/L as Se. Boron in the drainage water was reduced from 6–8 mg/L to 0.5 mg/L by an ion exchange post-treatment. This resin also removed residual selenium to below 10 mg/L. Trials on high-salinity drainage waters, similar to those found in evaporation ponds, were successful and gave enhanced specific selenium removal rates. The costs of removing selenium or selenium and boron from the drainage water were estimated to be $0.038–0.052/m^3 and $0.050–0.071/m^3, respectively, after allowance for byproduct recovery (boric acid and sodium sulfate) credits.

Squires, R.C. and R. Johnston. "Selenium Removal—Can We Afford It?" *Irrigation Systems for the 21st Century, Portland, Oregon, 1987.* **(American Society of Civil Engineers, New York, New York, pp. 455–467). 1987.**

The process of biologically removing the element selenium from agricultural drainage water is discussed, and an economic evaluation of the process is presented.

Steenhuis, T., R. Paulsen, T. Richard, W. Staubitz, M. Andreini and J. Surface. "Pesticide and Nitrate Movement under Conservation and Conventional Tilled Plots." *Planning Now for Irrigation and Drainage in the 21st Century, Lincoln, Nebraska, 1988.* **(American Society of Civil Engineers, New York, New York, pp. 587–595). 1988.**

Carbofuran, alachlor, atrazine, nitrate, and bromide (a tracer) were applied to plots with conventional and conservation tillage. Conventional tillage consisted of plowing, disking, and harrowing. In the conservation tilled plots, the sod was killed with Round-up® and the corn seeded without any further tillage.

During the early part of the growing season the conservation tilled plots had a higher tile discharge than those under conventional tillage due to dead sod cover that suppressed evapotranspiration. Low concentrations of atrazine and carbofuran were found below the root zone in the conservation tilled plots starting one month after application. In the conventional tillage it was not until late fall that some atrazine was detected below the root zone. Dye studies indicated that in the plowed layer of the conventional tilled plots water and solutes were in intimate contact with the soil matrix, promoting adsorption of the pesticides. The bromide tracer was not adsorbed and the bromide distribution with depth was similar for both tillage practices. Bromide was, therefore, a poor indicator for predicting potential pesticide losses under different tillage practices. Nitrate was only found in the zone that was never saturated.

Strutynski, B., R.E. Finger, S. Le and M. Lundt. "Pilot Scale Testing of Alternative Technologies for Meeting Effluent Reuse Criteria." *Water Environment Federation 65th Annual Conference & Exposition, New Orleans, Louisiana, 1992.* **(Water Environment Federation, Alexandria, Virginia, pp. 69–79). 1992.**

The Municipality of Metropolitan Seattle (Metro) has been investigating the potential for reuse of the effluent from its treatment plant at Renton for the last several years. The City of Seattle's Water Department, a major water supplier for the area, joined with Metro in 1991 in the evaluation and development of reuse options. These options include both nonconsumptive reuse, such as heating and cooling, and consumptive reuse, such as irrigation.

A pilot program was instituted in 1991 to identify a range of technologies capable of producing reuse quality water from secondary effluent. Technologies investigated included additional chlorination of the existing secondary effluent, as well as filtration, and ultraviolet disinfection without and with prior filtration. All process streams were monitored for a variety of parameters in addition to coliform levels. This paper presents the results of this testing.

Tim, U.S. and S. Mostaghimi. "Model for Predicting Virus Movement through Soils." *Ground Water.* **V. 29, n. 2. pp. 251–259. 1991.**

A numerical model, VIROTRANS, is developed for simulating the vertical movement of water and virus through soils treated with wastewater effluents and sewage sludges. The expression describing transient flow of water is coupled with the convective-dispersive equation for subsurface solute transport. The resulting methodology is a coupled set of partial differential equations that describe the transient flow of water and suspended virus particle movement through variably saturated media. Solutions to the partial differential equations are accomplished by a Galerkin finite element method. Several example problems are used to provide a quantitative verification and validation of the model. The model simulations are compared to an analytical solution and to experimental measurements of soil moisture content and poliovirus 1 transport.

The comparisons show reasonable agreement between model simulations and measured data. Sensitivity of the model's prediction to variations in pertinent input parameters are also analyzed.

Townley, J.A., S. Swanback and D. Andres. *Recharging a Potable Water Supply Aquifer with Reclaimed Wastewater in Cambria, California.* **John Carollo Engineers, Walnut Creek, CA. 1992.**

A proposed project in Cambria, a small unincorporated community in San Luis Obispo County, involves recharging one of the community's domestic supply aquifers with reclaimed wastewater. This paper describes the advance treatment system, regulatory involvement, and public acceptance issues.

Treweek, G.P. "Pretreatment Processes for Groundwater Recharge." *Artificial Recharge of Groundwater.* **Butterworth Publishers, Boston. pp. 205–248. 1985.**

Unplanned, indirect wastewater reuse through effluent discharge to streams and groundwater basins for subsequent downstream use by a wide variety of interests—agricultural, industrial, or domestic—has been a long-accepted practice throughout the world. Many communities at the end of major waterways, such as New Orleans and London, ingest water that already has been used as many as five times by repeated river withdrawal and discharge. Similarly, rivers or percolation basins may recharge underlying groundwater aquifers with reclaimed wastewater, which is in turn withdrawn by subsequent communities. For example, the effluent from over 140 wastewater treatment plants partially replenishes the groundwater tapped by the water supply system for London. This means effluent disposal, known as unplanned, indirect reuse, has become a generally accepted practice.

Planned, direct reuse is practiced on a smaller scale for a limited number of purposes, primarily agricultural and industrial. The terms "unplanned" and "planned" refer to whether the subsequent reuse was an unintentional byproduct of effluent discharge, or was designed as a conscious act following effluent discharge. The planned reuse schemes discussed in this report incorporated wastewater reclamation processes designed to meet not only effluent discharge standards, but also reuse standards promulgated by health authorities.

Troutman, D.E., E.M. Godsy, D.F. Goerlitz and G.G. Ehrlich. *Phenolic Contamination in the Sand-and-Gravel Aquifer from a Surface Impoundment of Wood Treatment Wastewaters, Pensacola, Florida.* **Geological Survey Water-Resources Investigations Report 84-4230. U.S. Geological Survey, Tallahassee, Florida. 1984.**

The discharge of creosote and pentachlorophenol wastewaters to unlined surface impoundments has resulted in groundwater contamination in the vicinity of a wood treatment plant near Pensacola, Florida. Total phenol concentrations of 36,000 µg/L have been detected at a depth 40 ft below land surface in a test hole 100 ft south of an overflow impoundment. Phenol concentrations in this same test hole were lest than 10 µg/L at a depth of 90 ft below land surface. Samples collected in test holes 1,350 ft downgradient from the surface impoundments and 100 ft north of Pensacola Bay, above and immediately below a clay lens, indicate that phenol-contaminated groundwater may not be discharging directly into Pensacola Bay. Phenol concentrations exceeding 20 µg/L were detected in samples from a drainage ditch discharging directly into Bayou Chico.

Microbiological data collected near the wood treatment site suggest that anaerobic methanogenic ecosystem contributes to reduction in phenol concentrations in groundwater. A laboratory study using bacteria isolated from the study site indicates that phenol, 2-methylphenol, and 3-methylphenol are significantly degraded and that methanogenesis reduces total phenol concentrations in laboratory digestors by 45%. Pentachlorophenol may inhibit methanogenesis at concentration exceeding 0.45 µg/L.

Data on wastewater migration in groundwater from American Creosote Works indicate that the sand-and-gravel aquifer is highly susceptible to contamination from unlined surface impoundments and other surface sources. Groundwater contamination occurs readily in pervious sands and gravel within the aquifer where the water table is near land surface. Coastal areas and valleys tend to be areas of groundwater discharge, and contamination of groundwater in these areas may result in surface water contamination.

U.S. Environmental Protection Agency Office of Water, Office of Wastewater Enforcement and Compliance, and Office of Research and Development, Office of Technology Transfer and Regulatory Support. *Manual: Guidelines for Water Reuse*. **EPA/625/R-92/004. U.S. Environmental Protection Agency, Washington, D.C. 1992.**

With many communities throughout the world approaching or reaching the limits of their available water supplies, water reclamation and reuse has become an attractive option for conserving and extending available water supplies. Water reuse may also present communities an opportunity for pollution abatement when it replaces effluent discharge to sensitive surface waters.

Water reclamation and nonpotable reuse only require conventional water and wastewater treatment technology that is widely practiced and readily available in countries throughout the world. Furthermore, because properly implemented nonpotable reuse does not entail significant health risks, it has generally been accepted and endorsed by the public in the urban and agricultural areas where it has been introduced.

Water reclamation for nonpotable reuse has been adopted in the United States and elsewhere without the benefit of national or international guidelines or standards. However, in recent years, many states in the U.S. have adopted standards or guidelines, and the World Health Organization (WHO) has published guidelines for reuse for agricultural irrigation. The primary purpose of this document is to present guidelines, with supporting information, for utilities and regulatory agencies in the U.S. In states where standards do not exist or are being revised or expanded, the guidelines can assist in developing reuse programs or appropriate regulations. The guidelines will also be useful to consulting engineers and others involved in the evaluation, planning, design, operation, or management of water reclamation and reuse facilities. In addition, a section on reuse internationally is offered to provide background and discuss relevant issues for authorities in other countries where reuse is being considered. The document does not propose standards by either the U.S. Environmental Protection Agency (EPA) or the U.S. Agency for International Development (AID). In the U.S., water reclamation and reuse standards are the responsibility of state agencies.

These guidelines primarily address water reclamation for nonpotable urban, industrial, and agricultural reuse, about which little controversy exists. Also, attention is given to augmentation of potable water supplies by indirect reuse. Because direct potable reuse is not currently practiced in the U.S., only a brief overview is provided.

Varuntanya, C.P. and D.R. Shafer. "Techniques for Fluoride Removal in Industrial Wastewaters." *Water Environment Federation 65th Annual Conference & Exposition, New Orleans, Louisiana, 1992.* **Water Environment Federation, Alexandria, Virginia, pp. 159–170. 1992.**

This paper will present data from several laboratory-scale treatability studies for fluoride removal from two industrial plant wastewaters. Additionally, limited plant data will be presented from one facility. Production at each facility involves the manufacture of zirconium tubes and the manufacture of semiconductors. The zirconium tube manufacturing plant wastewater contains approximately 15–20 mg/L fluoride before treatment, while the semiconductor facility contains approximately 25 mg/L fluoride. The results of the studies show that fluoride can be reduced to as low as 1 mg/L. The paper will also discuss the effluent concentrations achievable from each treatment scheme, as well as the chemical dosages required, and the process equipment necessary in each scheme. The advantages and disadvantages of the treatment processes will also be evaluated in this paper. The objective of the paper will provide insight for defluoridation of wastewaters for given effluent limitations.

Vaughn, J.M., E.F. Landry, L.J. Baranosky, C.A. Beckwith, M.C. Dahl and N.C. Delihas. "Survey of Human Virus Occurrence in Wastewater-Recharged Groundwater on Long Island." *Applied and Environmental Microbiology.* **V. 36, n. 1. pp. 47–51. 1978.**

Treated wastewater effluents and groundwater observation wells from three sewage recharge installations located on Long Island were assayed on a monthly basis for indigenous human enteroviruses and coliform bacteria for a period of 1 year. Viruses were detected in groundwater at sites where recharge basins were located less than 35 ft (ca. 10.6 m) above the aquifer. Results from one of the sites indicated the horizontal transfer of viable viruses through the groundwater aquifer.

Vecchioli, J., G.G. Ehrlich, E.M. Godsy and C.A. Pascale. "Alterations in the Chemistry of an Industrial Waste Liquid Injected into Limestone Near Pensacola, Florida." *Hydrogeology of Karstic Terrains: Case Histories.* **International Association of Hydrogeologists, UNESCO, V. 1. pp. 217–220. 1984.**

An industrial waste liquid containing organonitrile compounds and nitrate ions has been injected since June 1975 into the lower limestone of the Floridan aquifer at a site near Pensacola, Florida. Data from inorganic and organic chemicals, dissolved gas, and microbiological analyses of liquid backflowed from the injection well—and of liquid sample from a nearby monitor well—indicated that the injected waste liquid undergoes substantial chemical changes in the subsurface.

Verdin, J., G. Lyford and L. Sims. *Application of Satellite Remote Sensing for Identification of Irrigated Lands in the Newlands Project.* **1987.**

As one element of operating criteria and procedures for the Newlands project in west-central Nevada, the Bureau of Reclamation has compiled an irrigation water rights spatial database. In 1984, color infrared aerial photography was obtained and used to identify irrigated lands in the project. The photo interpretations were digitized to integrate them with water right maps for the project, which had similarly been digitized. Bench and bottomland designations, a soil type distinction of consequence for legal water entitlements, were recorded from maps as well. The database was used to calculate and summarize the tables, on a section-by-section basis, according to acreages of irrigated lands with water rights, irrigated lands without water rights, and nonirrigated water-righted land. For the 1985 and 1986 growing seasons, multispectral digital imagery of the project acquired by the Thermatic Mapper instrument on Landsat-5 was used to update the irrigated lands theme of the database. Scenes from May and August dates of those seasons, chosen after consideration of the phenologies of the major crops in the project, were coregistered and used to derive multidate vegetation index (Kauth-Thomas "greenness") images. These derivative images were then interpreted at an interactive video display providing a variety of enhancement capabilities, such as zooming and contrast stretching, to identify lands whose irrigation status had changed. Revisions to the irrigation theme of the spatial database were then made accordingly, as were modifications to the water rights coverage due to transfers between parcels of land. New acreage tabulations were prepared by digitally overlaying the revised coverages. In 1986, the SPOT-1 satellite was launched, and the higher-resolution imagery available from this remote sensing satellite is currently being evaluated for use in the Newlands Project. Multitemporal greenness images are being processed for crop type identification, and multispectral images are being digitally merged with 10-m resolution panchromatic images for improved interpretability.

Waller, B.G., B. Howie and C.R. Causaras. *Effluent Migration from Septic Tank Systems in Two Different Lithologies, Broward County, Florida.* **Geological Survey Water-Resources Investigations Report 87-4075. U.S. Geological Survey, Tallahassee, Florida. 1987.**

Two septic tank test sites, one in sand and one in limestone, in Broward County, Florida, were analyzed for effluent migration. Groundwater from shallow wells, both in background areas and hydraulically downgradient of the septic tank system, was sampled during a 16-month period from April 1983 through August 1984. Water quality indicators were used to determine the effluent-affected zone near the septic tank systems.

Specific conductance levels and concentrations of chloride, sulfate, ammonium, and nitrate indicated effluent movement primarily in a vertical direction with abrupt dilution as it moved downgradient. Effluent was detected in the sand to a depth more than 20 ft below the septic tank outlet, but was diluted to near background conditions 50 ft downgradient from the tank. Effluent in the limestone was detected in all three observation wells to depths exceeding 25 ft below the septic tank outlet and was diluted, but still detectable, 40 ft downgradient.

The primary controls on effluent movement from septic tank systems in Broward County are the lithology and layering of the geologic materials, hydraulic gradients, and the volume and type of use the system receives.

Wanielista, M., J. Charba, J. Dietz, R.S. Lott and B. Russell. *Evaluation of the Stormwater Treatment Facilities at the Lake Angel Detention Pond, Orange County, Florida.* **Report No. FL-ER-49-91. Florida Department of Transportation Environmental Office, Tallahassee, Florida. 1991.**

This is the final report on the use of Granulated Active Carbon (GAC) beds of Filtrasorb 400 in series to reduce the trihalomethane formation potential (THMFP) concentrations at the Lake Angel detention pond, Orange County, Florida. The detention pond accepts runoff from an interstate highway and a commercial area. Breakthrough time was estimated from laboratory analyses and used to design two beds in series at the detention pond. Breakthrough occurred in the first bed after treating 138,000 L of water. Exhaustion of the first bed was reached after treating 1270 bed volumes with a sorption zone length of 1.70 ft. The TOC adsorbed per gram of GAC was 6.3 mg. The liquid flow rate averaged 0.0011 cfs. Similar breakthrough curves for TOC and color were also reported. The used GAC can be disposed of by substituting it for sand in concrete mixes.

An economic evaluation of the GAC system at Lake Angel demonstrated an annual cost of $4.39/1000 gal to treat the stormwater runoff after detention and before discharge into a drainage well. This cost could be further reduced by using the stormwater to irrigate right-of-way sections of the watershed. An alternative method of pumping to another drainage basin was estimated to be more expensive.

The underdrain network for the GAC system initially became clogged with the iron- and sulfur-precipitating bacteria *Leptotrix*, *Gallionella* and *Thiothrix*. These bacteria were substantially reduced by altering the influent GAC system pipeline to take water directly from the lake. An alternative pipe system used a clay layer to reduce groundwater inputs and did not exhibit substantial bacterial growth.

Wellings, F.M. "Perspective on Risk of Waterborne Enteric Virus Infections." *Chemical and Biological Characterization of Sludges, Sediments, Dredge Spoils, and Drilling Muds. ASTM STP 976.* **American Society for Testing and Materials, Philadelphia, Pennsylvania. pp. 257–264. 1988.**

A valid perspective on the risk of waterborne enteric virus infections related to land disposal of sludge must incorporate various parameters. It is not enough to accept at face value the predominantly negative findings from the relatively few scientific studies that have purportedly been done to determine the fate of viruses introduced into the environment with the deposition of sludge. Rather, it is incumbent upon interested parties, whether they are regulatory or scientific, to evaluate the whole through a careful examination of the parts. This paper attempts to bring into focus the presently available data on four major virological issues. These are the characteristics of enteric viruses that enable them to survive wastewater treatment processes, problems associated with the accumulation and interpretation of data related to viral contamination of groundwater by sludge disposal practices, problems associated with transfer to the real world of data derived from laboratory-seeded experiments, and problems related to the establishment of the role of enteric viruses in waterborne disease outbreaks. Present data are insufficient for establishing the quantitative risk of waterborne disease due to the land disposal of sludge. However, there is some probability of groundwater contamination sufficient to warrant a cautious approach to land disposal of sludge.

White, E.M. and J.N. Dornbush. "Soil Changes Caused by Municipal Wastewater Applications in Eastern South Dakota." *Water Resources Bulletin*. V. 24, n. 2. pp. 269–273. 1988.

Wastewater from a municipal treatment plant was applied in rapid infiltration basins for four years to determine a poorly drained soil's effectiveness in removing influent nitrogen and phosphorus and the soil changes that might limit their removal. About half the total PO_4-P lost from the influent was sorbed in the upper 91 cm of the soil, and the other half was sorbed by the soil below the perforated pipe, which was used to drain the basins and collect the effluent for analysis. Drying of the basin soils converted more sorbed PO_4-P to Ca phosphates, but the total sorbed was about the same. The influent nitrogen decreased, probably by volatilization, because the two basins with surface soil lost soil nitrogen rather than gained soil nitrogen. The soil total Ca, Mg, and K contents did not change significantly but nitrogen increased slightly. Changes in the characteristics of the soils were slight, and would have little effect on the longevity of a rapid infiltration basin.

Wilde, F.D. *Geochemistry and Factors Affecting Ground-Water Quality at Three Storm-Water-Management Sites in Maryland: Report of Investigations No. 59*. Department of Natural Resources, Maryland Geological Survey, Baltimore, Maryland. (Prepared in cooperation with the U.S. Department of the Interior Geological Survey, The Maryland Department of the Environment, and The Governor's Commission on Chesapeake Bay Initiatives). 1994.

The effects of infiltration of storm runoff on groundwater chemistry and quality were examined at three suburban stormwater management impoundments at sites in Annapolis, Greenmount, and Prince Frederick, Maryland. Geochemical and hydrologic data were collected from December 1985 through June 1989. The Annapolis and Prince Frederick sites are in the Coastal Plain physiographic province, and the Greenmount site is in the Piedmont physiographic province. This study was a cooperative effort of the U.S. Geological Survey, the Maryland Geological Survey, the Maryland Department of the Environment, and the Governor's Council on Chesapeake Bay Initiatives.

The objectives of the study were to determine whether the chemical composition of groundwater beneath the impoundments changed as a result of stormwater infiltration, whether groundwater quality was adversely affected, and whether contaminants being sequestered in impoundments could be potentially mobilized to groundwater.

Native geographic materials collected from drill cores at each site and bottom materials collected each September from the Annapolis and Greenmount ponds were analyzed for particle size, selected chemical constituents, and the predominant mineralogy. Aqueous solutions were analyzed for major ions, a large suite of trace elements, and volatile and polyaromatic organic compounds; pH, dissolved oxygen, alkalinity, chloride, and specific conductance were measured in the field. Samples of runoff, impoundment water, unsaturated zone water, and groundwater were collected triannually for extensive chemical analyses and at least monthly for field measurements. The results were compared for the two sites with stormwater ponds (Annapolis and Greenmount) and a porous-pavement site with a subsurface impoundment (Prince Frederick).

Either primary or secondary maximum contaminant levels established by the U.S. Environmental Protection Agency (USEPA) for aluminum, cadmium, chloride, chromium, and lead in drinking water were exceeded from time to time in groundwater samples collected beneath and downgradient from the impoundments. In addition, uncharacteristically high concentrations of barium, copper, molybdenum, nickel, strontium, vanadium, and zinc occasionally were reported in groundwater beneath the impoundments, and median concentrations of barium, cadmium, chloride, copper, nickel, and zinc were elevated in some groundwater samples beneath the study sites. Chromium and lead were rarely detected in groundwater. Low concentrations of arsenic were detected sporadically in stormwater and groundwater. Concentrations of volatile organic compounds were usually below or near detection in stormwater and groundwater samples; small concentrations of polyaromatic organic compounds were detected only in pond bottom materials.

Pond bottom materials generally were effective scavengers of trace metals introduced to stormwater impoundments in the runoff. Between 1986 through 1988, concentrations of lead increased from below detection levels to 28 ppm, and zinc concentrations increased from 54 to 344 ppm in bottom materials collected from the Annapolis impoundment. In addition, copper concentrations increased from 3.5 to 40 ppm, and nickel concentrations increased from 6 to 16 ppm in bottom materials at the Annapolis site. For the same period at the Greenmount impoundment, concentrations of lead increased from 20 to 90 ppm; zinc concentrations increased from 59 to 469 ppm; and nickel concentrations increased from 35 to 48 ppm. (There was a net decrease in the copper concentration.) Despite this accumulation of metals in bottom materials, concentrations of these and other metals were considerably elevated in groundwater beneath and downgradient from the impoundments. Cadmium concentrations did not increase in bottom materials, although cadmium was a common constituent in stormwater and sorbed readily to the bottom materials in laboratory tests. Groundwater samples collected from the control wells at each site had concentrations of cadmium that were below or near detection, whereas concentrations were as high as 27, 26, and 8.4 µg/L beneath the impoundments at Annapolis, Greenmount, and Prince Frederick, respectively.

Stormwater was the primary source of most contaminants that were found at elevated concentrations in groundwater samples collected beneath impoundments. Contaminants entered groundwater as a result of several variables, including direct stormwater infiltration; impoundment-related modifications of pH and redox that periodically favored metal mobilization from pond bottom or aquifer materials; and formation of anionic or neutral complexes.

Metal mobility in the impoundments was mitigated by ion exchange, sorption, and mineral precipitation; stormwater was aerobic and usually had neutral or higher pH that generally did not favor the presence of soluble species of most metals. Nevertheless, cadmium sorption in the impoundments may have been excluded by competing cations. Moreover, conditions for complexing of cadmium with organic compounds and chloride was favorable in impoundments, and could have enhanced cadmium transport to groundwater.

Algal photosynthesis modified the pond water chemistry at Annapolis and Greenmount, increasing pH to 9.0 or greater, whereas algal respiration and rainwater dilution decreased pH to 6.5. Algal mediation of the pond water pH at Greenmount resulted in a median pH of 9.2. Because aluminum solubility increases exponentially at about pH 9.0, aluminum concentrations in pond water at this site exceeded the U.S. Environmental Protection Agency's drinking water regulation of 50 µg/L in most samples and may have contributed to elevated aluminum concentrations measured in groundwater beneath the impoundment. Algal activity was less intense in the Annapolis impoundment, where the median pond water pH was 7.7 and the median aluminum concentration was 30 µg/L; occasional measurements of aluminum concentrations greater than 50 µg/L in pond water corresponded with pH near or greater than 9.0. In groundwater beneath the Annapolis pond, aluminum concentrations were below 50 µg/L in the samples collected, with one exception.

The solubility of most trace metals increases with pH decrease below neutrality. Therefore, the decreases in pond pH below neutral possibly mobilized iron, copper, nickel, and zinc periodically from bottom materials; alternatively, metals dissolved in stormwater could have been transported to the groundwater because kinetics were unfavorable for sorption to bottom materials.

The pH of groundwater tended to keep the metals in solution at each site. Groundwater pH beneath the impoundments was reduced to below background pH; from 5.13 to between 4.18 and 4.94 at Annapolis, from 5.38 to between 4.9 and 5.29 at Greenmount, and from 6.71 to between 4.39 and 4.8 at Prince Frederick. Periodic mobilization of iron from the impoundment to the groundwater (and the consequent precipitation of iron hydroxides in the aquifer) has been suggested as a cause of reduced pH in beneath pond groundwater at the Annapolis and Greenmount sites. Lithologic composition is the primary control on order-of-magnitude changes in groundwater pH at Prince Frederick.

With the exception of chloride, groundwater contamination was least at the Prince Frederick site, possibly because of low contaminant concentrations in stormwater entering the Prince Frederick impoundment, or because contaminant mobility was restricted by the well buffered and stable chemical environment in the impoundment. Dissolution of the rock aggregate in the Prince Frederick impoundment buffered impounded water to a pH of about 8.4, but dissolution also released high concentrations of magnesium and low concentrations of nickel and possibly chromium that affected groundwater chemistry.

Chloride contamination was ubiquitous in groundwater receiving storm runoff infiltrate. The groundwater beneath the impoundments was modified to a chloride-dominated solution at each site throughout the period of study. (Native groundwaters were mixed cation/mixed anion, magnesium-nitrate/chloride, and calcium-bicarbonate types at respective study sites.) The only major source of chloride to groundwater was stormwater infiltration during periods of road salting (road salting occurred no more than five times a year during the study period).

Chloride concentrations were measured at least biweekly and before, during, and after selected storms. The effect of seasonal and storm-specific recharge on concentrations of chloride in groundwater beneath the impoundments monitored was a temporary dilution—usually lasting no longer than several days. Chloride concentrations in groundwater beneath impoundments increased whenever rainfall was low and evapotranspiration rates were high, resulting in highest concentrations during the summer and early autumn at each site. The magnitude and persistence of the chloride contamination indicated that chloride, although a very soluble constituent, was not being flushed readily from the groundwater systems studied. Chloride concentrations in groundwater beneath impoundments increased whenever rainfall was low and evapotranspiration rates were high.

Probable factors contributing to the persistent chloride domination of the major-ion chemistry of groundwater at each study site were (1) low groundwater flow rates relative to stormwater infiltration rates; (2) limited dilution potential because sites were within a maximum of 300 ft from the inferred groundwater divide; and (3) capillary forces.

The unsaturated-zone chloride data suggest that capillary processes cause retardation of chloride transport and can allow chloride buildup, especially in the zone of tension saturation. This could serve as a model for explaining inhibition of transport of other contaminants. With the exception of chloride concentrations, however, the periods of data collection and/or sampling frequency generally were insufficient to determine temporal trends in concentrations for trace metals and other constituents.

Wilson, L.G., M.D. Osborn, K.L. Olson, S.M. Maida and L.T. Katz. "The Ground Water Recharge and Pollution Potential of Dry Wells in Pima County, Arizona." *Ground Water Monitoring Review*. V. 10. pp. 114–121. 1990.

This paper summarizes a study to estimate the potential for dry well drainage of urban runoff to recharge and pollute groundwater in Tucson, Arizona. We selected three candidate dry wells for study. At each site we collected samples of runoff, dry well sediment, vadose zone sediment, perched groundwater, and groundwater. Water content data from vadose zone samples suggest that dry well drainage has created a transmission zone for water movement at each site. Volatile organic compounds, while undetected in runoff samples, were present in dry well sediment, perched groundwater at one site, and groundwater at two sites. The concentrations of volatile organics (toluene and ethylbenzene) in the water samples were less than the corresponding EPA human health criteria. Pesticides were detected only in runoff and dry well sediment. Lead and chromium occurred in runoff samples at concentrations above drinking water standards. Nickel, chromium, and zinc concentrations were elevated in vadose zone samples at the commercial site. Of the metals, only manganese, detected at the residential site, exceeded Secondary Drinking Water Standards in groundwater. It is concluded that the three dry wells examined during this study are currently not a major source of groundwater pollution.

Wolff, J., J. Ebeling, A. Muller and H. Wacker. "Waste Water Irrigation Suited to the Environment as Shown by the Example of the 'Abwasserverband Wolfsburg'." *Hydrological Process and Water Management in Urban Areas, Duisburg, Federal Republic of Germany, 1988.* **(International Hydrological Programme, UNESCO, pp. 599–606). 1988.**

Changes in the chemistry of groundwater by the influences of wastewater land treatment will be discussed by the research at the area of the 'Abwasserverband Wolfsburg,' southeast Lower Saxony. The results of a four-year research programme show that, under certain conditions, wastewater land treatment can be realized in an ecologically justifiable way.

Yurewicz, M.C. and J.C. Rosenau. *Effects on Ground Water of Spray Irrigation Using Treated Municipal Sewage Southwest of Tallahassee, Florida.* **Geological Survey Water-Resources Investigation Report 86-4109. U.S. Geological Survey, Tallahassee, Florida. 1986.**

Increases in the concentrations of chloride and nitrate-nitrogen in groundwater have resulted from land application of secondary-treated municipal sewage southwest of Tallahassee, Florida. The increases occurred predominantly during periods of above normal application rates. This result is based upon a data collection program which began in 1972, 6 years after the initial application of treated sewage. The data collection period for this report is 1982 through June 1981.

Although an estimated minimum volume of 4,220 million gallons of treated sewage was spray-irrigated from July 1966 through June 1981, distortion of the local groundwater flow pattern did not occur because of the high natural recharge and high permeability of the limestone aquifer. Direct recharge from the land surface to the Floridan aquifer system occurs by rapid infiltration through the sand overburden and a discontinuous clay layer above the limestone formation. Soluble constituents move laterally and vertically with the groundwater flow pattern. Use of chloride as a tracer of water movement indicates that treated sewage occurs at depths greater than 200 ft below land surface below the spray sites. The direction and rate of groundwater movement is southwesterly toward the Gulf of Mexico, at a rate of approximately 5 ft/day, with significant downward movement also occurring.

The most significant effect on groundwater quality has been high nitrate-nitrogen concentrations, which were detected between 1972 and 1976 when high volumes of treated sewage were applied for experimental purposes. During this period, nitrate-nitrogen concentrations in the upper limestones of the Floridan aquifer system exceeded the maximum contaminant level of 10 mg/L established for potable water supplies. Computations indicate that if the monthly load of nitrogen does not exceed 130 to 180 lb/ac, the concentration of nitrate-nitrogen in the upper part of the aquifer will not exceed 10 mg/L.

Other water quality characteristics were not significantly affected by the application of treated sewage. Concentrations of trace metals—including arsenic, cadmium, chromium, copper, iron, lead, manganese, mercury, selenium, and zinc—in groundwater remained at background levels. Organochlorine insecticides and chlorinated phenoxy acid herbicides were analyzed for, but not detected, in 18 groundwater samples collected in 1974 and 1978. Concentrations of major inorganic ions in the groundwater likely are controlled by equilibrium conditions between the water and the aquifer matrix.

GROUNDWATER CONTAMINATION BIBLIOGRAPHY

Alhajjar, B.J., G.V. Simsiman and G. Chesters. "Fate and Transport of Alachlor, Metolachlor and Atrazine in Large Columns." *Water Science and Technology: A Journal of the International Association of Water Pollution Research.* Vol. 22, No. 6. pp. 87–94. 1990.

American Water Works Association. "Fertilizer Contaminates Nebraska Groundwater." *AWWA Mainstream.* Vol. 34, No. 4. pp. 6. 1990.

Amoros, I., J.L. Alonso and I. Peris. "Study of Microbial Quality and Toxicity of Effluents from Two Treatment Plants Used for Irrigation." *Water Science and Technology: A Journal of the International Association on Water Pollution Research.* Vol. 21, No. 3. pp. 243–246. 1989.

Armstrong, D.E. and R. Llena. *Stormwater Infiltration: Potential for Pollutant Removal.* Project Report to the U.S. Environmental Protection Agency. Wisconsin Department of Natural Resources, Water Chemistry Program. 1992.

Aronson, D.A. and G.E. Seaburn. *Appraisal of Operating Efficiency of Recharge Basins on Long Island, New York, in 1969.* Geological Survey Water-Supply Paper 2001-D. U.S. Government Printing Office, Washington, D.C. 1974.

Asano, T. "Irrigation with Reclaimed Wastewater—Recent Trends." *Irrigation Systems for the 21st Century, Portland, Oregon, 1987.* (American Society of Civil Engineers, New York, New York, pp. 735–742). 1987.

Bao-rui, Y. "Investigation into Mechanisms of Microbial Effects on Iron and Manganese Transformations in Artificially Recharged Groundwater." *Water Science and Technology: A Journal of the International Association on Water Pollution Research.* Vol. 20, No. 3. pp. 47–53. 1988.

Barraclough, J.T. "Waste Injection into a Deep Limestone in Northwestern Florida." *Ground Water.* Vol. 3, No. 1. pp. 22–24. 1966.

Boggess, D.H. *Effects of a Landfill on Ground Water Quality.* Geological Survey Open File Report 75-594. U.S. Geological Survey, Tallahassee, Florida. 1975.

Bouchard, A.B. "Virus Inactivation Studies for a California Wastewater Reclamation Plant." *Water Environment Federation 65th Annual Conference & Exposition, New Orleans, Louisiana, 1992.* (Water Environment Federation, Alexandria, Virginia). 1992.

Bouwer, E.J., P.L. McCarty, H. Bouwer and R.C. Rice. "Organic Contaminant Behavior during Rapid Infiltration of Secondary Wastewater at the Phoenix 23rd Avenue Project." *Water Resource.* Vol. 18, No. 4. pp. 463–472. 1984.

Bouwer, H. "Agricultural Contamination: Problems and Solutions." *Water Environment and Technology.* Vol. 1, No. 2. pp. 292–297. 1989.

Bouwer, H. "Effect of Irrigated Agriculture on Groundwater." *Journal of Irrigation and Drainage Engineering.* Vol. 113, No. 1. pp. 4–15. 1987.

Bouwer, H. and E. Idelovitch. "Quality Requirements for Irrigation with Sewage Water," *Journal of Irrigation and Drainage Engineering.* Vol. 113, No. 4. pp. 516–535. 1987.

Brown, D.P. *Effects of Effluent Spray Irrigation on Ground Water at a Test Site Near Tarpon Springs, Florida.* Geological Survey Open-File Report 81-1197. U.S. Geological Survey, Tallahassee, Florida. 1982.

Butler, K.S. "Urban Growth Management and Groundwater Protection: Austin, Texas." *Planning for Groundwater Protection.* Academic Press, New York, New York. pp. 261–287. 1987.

Cavanaugh, J.E., H.S. Weinberg, A. Gold, R. Sangalah, D. Marbury, W.H. Glaze, T.W. Collette, S.D. Richardson and A.D. Thurston Jr. "Ozonation Byproducts: Identification of Bromohydrins from the Ozonation of Natural Water with Enhanced Bromide Levels." *Environmental Science & Technology.* Vol. 26, No. 8. pp. 1658–1662. 1992.

Chang, A.C., A.L. Page, P.F. Pratt and J.E. Warneke. "Leaching of Nitrate from Freely Drained-Irrigated Fields Treated with Municipal Sludges." *Planning Now for Irrigation and Drainage in the 21st Century, Lincoln, Nebraska, 1988.* (American Society of Civil Engineers, New York, New York, pp. 455–467). 1988.

Chase, W.L.J. "Reclaiming Wastewater in Phoenix, Arizona." *Irrigation Systems for the 21st Century, Portland, Oregon, 1987.* (American Society of Civil Engineers, New York, New York, pp. 336–343). 1987.

Cisic, S., D. Marske, B. Sheikh, B. Smith and F. Grant. "City of Los Angeles Effluent—Today's Waste, Tomorrow's Resource." *Water Environment Federation 65th Annual Conference & Exposition, New Orleans, Louisiana, 1992.* (Water Environment Federation, Alexandria, Virginia). 1992.

Close, M.E. "Effects of Irrigation on Water Quality of a Shallow Unconfined Aquifer." *Water Resources Bulletin.* Vol. 23, No. 5. pp. 793–802. 1987.

Clothier, B.E. and T.J. Sauer. "Nitrogen Transport during Drip Fertigation with Urea." *Soil Science Society of America Journal.* Vol. 52, No. 2. pp. 345–349. 1988.

Craun, G.F. "Waterborne Disease—A Status Report Emphasizing Outbreaks In Ground-Water Systems." *Ground Water.* Vol. 17, No. 2. pp. 183–191. 1979.

Crites, R.W. "Micropollutant Removal in Rapid Infiltration." *Artificial Recharge of Groundwater.* Butterworth Publishers, Boston. pp. 579–608. 1985.

Crook, J., T. Asano and M. Nellor. "Groundwater Recharge with Reclaimed Water in California." *Water Environment & Technology.* Vol. 2, No. 8. pp. 42–49. 1990.

Deason, J.P. "Irrigation-Induced Contamination: How Real a Problem?" *Journal of Irrigation and Drainage Engineering.* Vol. 115, No. 1. pp. 9–20. 1989.

Deason, J.P. "Selenium: It's Not Just in California." *Irrigation Systems for the 21st Century, Portland, Oregon, 1987.* (American Society of Civil Engineers, New York, New York, pp. 475–482). 1987.

DeBoer, J.G. "Wastewater Reuse: A Resource or a Nuisance?" *Journal of the American Water Works Association.* Vol. 75. pp. 348–356. 1983.

Domagalski, J.L. and N.M. Dubrovsky. "Pesticide Residues in Ground Water of the San Joaquin Valley, California." *Journal of Hydrology.* Vol. 130, No. 1–4. pp. 299–338. 1992.

Ehrlich, G.G., E.M. Godsy, C.A. Pascale and J. Vecchioli. "Chemical Changes in an Industrial Waste Liquid during Post-Injection Movement in a Limestone Aquifer, Pensacola, Florida." *Ground Water.* Vol. 17, No. 6. pp. 562–573. 1979.

Ehrlich, G.G., H.F.H. Ku, J. Vecchioli and T.A. Ehlke. *Microbiological Effects of Recharging the Magothy Aquifer, Bay Park, New York, with Tertiary-Treated Sewage.* Geological Survey Professional Paper 751-E. U.S. Government Printing Office, Washington, D.C. 1979.

Elder, J.F., J.D. Hunn and C.W. Calhoun. *Wastewater Application by Spray Irrigation on a Field Southeast of Tallahassee, Florida: Effects on Ground-Water Quality and Quantity.* Geological Survey Water-Resources Investigations Report 85-4006. U.S. Geological Survey, Tallahassee, Florida. 1985.

Eren, J. "Changes in Wastewater Quality during Long Term Storage." Mekoroth Water Co., Northern District, P.O.B. 755, Haifa, Israel. pp. 1291–1300.

Ferguson, B.K. "Role of the Long-Term Water Balance in Management of Stormwater Infiltration." *Journal of Environmental Management.* Vol. 30, No. 3. pp. 221–233. 1990.

Ferguson, R.B., D.E. Eisenhauer, T.L. Bockstadter, D.H. Krull and G. Buttermore. "Water and Nitrogen Management in Central Platte Valley of Nebraska." *Journal of Irrigation and Drainage Engineering.* Vol. 116, No. 4. pp. 557–565. 1990.

Gerba, C.P. and S.M. Goyal. "Pathogen Removal from Wastewater during Groundwater Recharge." *Artificial Recharge of Groundwater.* Butterworth Publishers, Boston. pp. 283–317. 1985.

Gerba, C.P. and C.N. Haas. "Assessment of Risks Associated with Enteric Viruses in Contaminated Drinking Water." *Chemical and Biological Characterization of Sludges, Sediments, Dredge Spoils, and Drilling Muds. ASTM STP 976.* American Society for Testing and Materials, Philadelphia, Pennsylvania. pp. 489–494. 1988.

German, E.R. *Quantity and Quality of Stormwater Runoff Recharged to the Floridan Aquifer System Through Two Drainage Wells in the Orlando, Florida, Area.* Geological Survey Water Supply Paper 2344. U.S. Government Printing Office, Washington, D.C.. 1989.

Goldschmid, J. "Water-Quality Aspects of Ground-Water Recharge in Israel." *Journal of the American Water Works Association.* Vol. 66, No. 3. pp. 163–166. 1974.

Goolsby, D.A. *Geochemical Effects and Movement of Injected Industrial Waste in a Limestone Aquifer.* Memoir No. 18. American Association of Petroleum Geologists. 1972.

Greene, G.E. *Ozone Disinfection and Treatment of Urban Storm Drain Dry-Weather Flows: A Pilot Treatment Plant Demonstration Project on the Kenter Canyon Storm Drain System in Santa Monica.* The Santa Monica Bay Restoration Project, Monterey Park, CA. 1992.

Hampson, P.S. *Effects of Detention on Water Quality of Two Stormwater Detention Ponds Receiving Highway Surface Runoff in Jacksonville, Florida.* Geological Survey Water-Resources Investigations Report 86-4151. U.S. Geological Survey, Tallahassee, Florida. 1986.

Harper, H.H. *Effects of Stormwater Management Systems on Groundwater Quality.* DER Project WM190. Florida Department of Environmental Regulation, Orlando, Florida. 1988.

Hickey, J.J. "Subsurface Injection of Treated Sewage into a Saline-Water Aquifer at St. Petersburg, Florida—Aquifer Pressure Buildup." *Ground Water.* Vol. 22, No. 1. pp. 48–55. 1984.

Hickey, J.J. and J. Vecchioli. *Subsurface Injection of Liquid Waste with Emphasis on Injection Practices in Florida.* Geological Survey Water-Supply Paper 2281. United States Government Printing Office, Washington D.C. 1986.

Hickey, J.J. and W.E. Wilson. *Results of Deep-Well Injection Testing at Mulberry, Florida.* Geological Survey Water-Resources Investigations Report 81-75. U.S. Geological Survey, Tallahassee, Florida. 1982.

Higgins, A.J. "Impacts on Groundwater Due to Land Application of Sewage Sludge." *Water Resources Bulletin*. Vol. 20, No. 3. pp. 425–434. 1984.

Horsley, S.W. and J.A. Moser. "Monitoring Ground Water for Pesticides at a Golf Course—A Case Study on Cape Cod, Massachusetts." *Ground Water Monitoring Review*. Vol. 10. pp. 101–108. 1990.

Hull, R.W. and M.C. Yurewicz. *Quality of Storm Runoff to Drainage Wells in Live Oak, Florida, April 4, 1979*. Geological Survey Open-File Report 79-1073. U.S. Government Printing Office, Washington, D.C. 1979.

Ishizaki, K. "Control of Surface Runoff by Subsurface Infiltration of Stormwater: A Case Study in Japan." *Artificial Recharge of Groundwater*. Butterworth Publishers, Boston. pp. 565–575. 1985.

Jansons, J., L.W. Edmonds, B. Speight and M.R. Bucens. "Movement of Viruses after Artificial Recharge." *Water Research*. Vol. 23, No. 3. pp. 293–299. 1989.

Jansons, J., L.W. Edmonds, B. Speight and M.R. Bucens. "Survival of Viruses in Groundwater." *Water Research*. Vol. 23, No. 3. pp. 301–306. 1989.

Johnson, R.B. "The Reclaimed Water Delivery System and Reuse Program for Tucson, Arizona." *Irrigation Systems for the 21st Century, Portland, Oregon, 1987*. (American Society of Civil Engineers, New York, New York, pp. 344–351). 1987.

Karkal, S.S. and D.L. Stringfield. "Wastewater Reclamation and Small Communities: A Case History." *Water Environment Federation 65th Annual Conference & Exposition, New Orleans, Louisiana, 1992*. (Water Environment Federation, Alexandria, Virginia, pp. 419–425). 1992.

Katopodes, N.D. and J.H. Tang. "Self-Adaptive Control of Surface Irrigation Advance." *Journal of Irrigation and Drainage Engineering*. Vol. 116, No. 5. pp. 696–713. 1990.

Kaufman, M.I. "Subsurface Wastewater Injection, Florida." *Journal of Irrigation and Drainage Engineering*. Vol. 99, No. 1. pp. 53–70. 1973.

Knisel, W.G. and R.A. Leonard. "Irrigation Impact on Groundwater: Model Study in Humid Region," *Journal of Irrigation and Drainage Engineering*. Vol. 115, No. 5. pp. 823–839. 1989.

Krawchuk, B.P. and B.G.R. Webster. "Movement of Pesticides to Ground Water in an Irrigated Soil." *Water Pollution Research Journal of Canada*. Vol. 22, No. 1. pp. 129–146. 1987.

Ku, H.F.H., N.W. Hagelin and H.T. Buxton. "Effects of Urban Storm-Runoff Control on Ground-Water Recharge in Nassau County, New York." *Ground Water*. Vol. 30, No. 4. pp. 507–513. 1992.

Ku, H.F.H. and D.L. Simmons. *Effect of Urban Stormwater Runoff on Ground Water beneath Recharge Basins on Long Island, New York*. Geological Survey Water-Resources Investigations Report 85-4088. U.S. Geological Survey, Syosset, New York. 1986.

Ku, H.F.H., J. Vecchioli and S.E. Ragone. "Changes in Concentration of Certain Constituents of Treated Waste Water during Movement through the Magothy Aquifer, Bay Park, New York." *Journal Research U.S. Geology Survey*. Vol. 3, No. 1. pp. 89–92. 1975.

Lauer, D.A. "Vertical Distribution in Soil of Sprinkler-Applied Phosphorus." *Soil Science Society of America Journal*. Vol. 52, No. 3. pp. 862–868. 1988.

Lauer, D.A. "Vertical Distribution in Soil of Unincorporated Surface-Applied Phosphorus under Sprinkler Irrigation." *Soil Science Society of America Journal*. Vol. 52, No. 6. pp. 1685–1692. 1988.

Lee, E.W. "Drainage Water Treatment and Disposal Options." *Agricultural Salinity Assessment and Management*. American Society of Civil Engineers. pp. 450–468. 1990.

Lloyd, J.W., D.N. Lerner, M.O. Rivett and M. Ford. "Quantity and Quality of Groundwater Beneath an Industrial Conurbation—Birmingham, UK." *Hydrological Processes and Water Management in Urban Areas, Duisburg, Federal Republic of Germany, 1988*. (International Hydrological Programme, UNESCO, pp. 445–453). 1988.

Loh, P.C., R.S. Fujioka and W.M. Hirano. "Thermal Inactivation of Human Enteric Viruses in Sewage Sludge and Virus Detection by Nitrose Cellulose-Enzyme Immunoassay." *Chemical and Biological Characterization of Sludges, Sediments, Dredge Spoils, and Drilling Muds. ASTM STP 976.* American Society for Testing and Materials, Philadelphia, Pennsylvania. pp. 273–281. 1988.

Malik, A., M. Stone, F.R. Martinez and R. Paul. "First Wastewater Desalting Plant in Central Coast, California." *Water Environment Federation 65th Annual Conference & Exposition, New Orleans, Louisiana, 1992.* (Water Environment Federation, Alexandria, Virginia, pp. 395–406). 1992.

Mancini, J.L. and A.H. Plummer Jr. "A Method for Developing Wet Weather Water Quality Criteria for Toxics." *Water Environment Federation 65th Annual Conference & Exposition, New Orleans, Louisiana, 1992.* (Water Environment Federation, Alexandria, Virginia, pp. 15–26). 1992.

Markwood, I.M. "Waterborne Disease—Historical Lesson." *Ground Water.* Vol. 17, No. 2. pp. 197–198. 1979.

Marton J. and Mohler I. "The Influence of Urbanization on the Quality of Groundwater." *Hydrological Processes and Water Management in Urban Areas, Duisburg, Federal Republic of Germany, 1988.* (International Hydrological Programme, UNESCO, pp. 452–461). 1988.

Marzouk, Y., S.M. Goyal and C.P. Gerba. "Prevalence of Enteroviruses in Ground Water of Israel." *Ground Water.* Vol. 17, No. 5. pp. 487–491. 1979.

Merkel, B., J. Grossmann and P. Udluft. "Effect of Urbanization on a Shallow Quaternary Aquifer." *Hydrological Processes and Water Management in Urban Areas, Duisburg, Federal Republic of Germany, 1988.* (International Hydrological Programme, UNESCO, pp. 461–469). 1988.

Mossbarger, W.A. Jr. and R.W. Yost. "Effects of Irrigated Agriculture on Groundwater Quality in Corn Belt and Lake States." *Journal of Irrigation and Drainage Engineering.* Vol. 115, No. 5. pp. 773–789. 1990.

Nellor, M.H., R.B. Baird and J.R. Smyth. "Health Aspects of Groundwater Recharge." *Artificial Recharge of Groundwater.* Butterworth Publishers, Boston. pp. 329–355. 1985.

Nightingale, H.I., J.E. Ayars, R.L. McCormick and D.C. Cehrs. "Leaky Acres Recharge Facility: A Ten-Year Evaluation." *Water Resources Bulletin.* Vol. 19, No. 3. pp. 429–437. 1983.

Nightingale, H.I. "Accumulation of As, Ni, Cu, and Pb in Retention and Recharge Basins Soils from Urban Runoff." *Water Resources Bulletin.* Vol. 23, No. 4. pp. 663–672. 1987.

Nightingale, H.I. "Water Quality beneath Urban Runoff Management Basins." *Water Resources Bulletin.* V. 23, No. 2. pp. 197–205. 1987.

Nightingale, H.I. and W.C. Bianchi. "Ground-Water Chemical Quality Management by Artificial Recharge." *Ground Water.* V. 15, No. 1. pp. 15–22. 1977.

Nightingale, H.I. and W.C. Bianchi. "Ground-Water Turbidity Resulting from Artificial Recharge." *Ground Water.* Vol. 15, No. 2. pp. 146–152. 1977.

Norberg-King, T.J., E.J. Durhan, G.T. Ankley and E. Robert. "Application of Toxicity Identification Evaluation Procedures to the Ambient Waters of the Colusa Basin Drain, California." *Environmental Toxicology and Chemistry.* Vol. 10. pp. 891–900. 1991.

Pahren, H.R. "EPA's Research Program on Health Effects of Wastewater Reuse for Potable Purposes." *Artificial Recharge of Groundwater.* Butterworth Publishers, Boston. pp. 319–328. 1985.

Peterson, D.A. "Selenium in the Kendrick Reclamation Project, Wyoming." *Planning Now for Irrigation and Drainage in the 21st Century, Lincoln, Nebraska, 1988.* (American Society of Civil Engineers, New York, New York, pp. 678–685). 1988.

Petrovic, A.M. "The Fate of Nitrogenous Fertilizers Applied to Turfgrass." *Journal of Environmental Quality.* Vol. 19, No. 1. pp. 1–14. 1990.

Phelps, G.G. *Effects of Surface Runoff and Treated Wastewater Recharge on Quality of Water in the Floridan Aquifer System, Gainesville Area, Alachua County, Florida.* Geological Survey Water-Resources Investigations Report 87-4099. U.S. Geological Survey, Tallahassee, Florida. 1987.

Pierce, R.C. and M.P. Wong. "Pesticides in Agricultural Waters: The Role of Water Quality Guidelines." *Canadian Water Resources Journal.* Vol. 13, No. 3. pp. 33–49. 1988.

Pitt, W.A.J. Jr. *Effects of Septic Tank Effluent on Ground-Water Quality, Dade County, Florida: An Interim Report.* Geological Survey Open File Report 74010. U.S. Geological Survey, Tallahassee, Florida. 1974.

Pitt, W.A.J. Jr., H.C. Mattraw and H. Klein. *Ground-Water Quality in Selected Areas Serviced by Septic Tanks, Dade County, Florida.* Geological Survey Open File Report 75-607. U.S. Geological Survey, Tallahassee, Florida. 1975.

Power, J.F. and J.S. Schepers. "Nitrate Contamination of Groundwater in North America." *Agriculture, Ecosystems and Environment.* Vol. 26, No. 3-4. pp. 165–187. 1989.

Pruitt, J.B., D.E. Troutman and G.A. Irwin. *Reconnaissance of Selected Organic Contaminants in Effluent and Ground Water at Fifteen Municipal Wastewater Treatment Plants in Florida.* Geological Survey Water-Resources Investigations Report 85-4167. U.S. Geological Survey, Tallahassee, Florida. 1985.

Ragone, S.E. *Geochemical Effects of Recharging the Magothy Aquifer, Bay Park, New York, with Tertiary-Treated Sewage.* Geological Survey Professional Paper 751-D. U.S. Government Printing Office, Washington, D.C. 1977.

Ragone, S.E., H.F.H. Ku and J. Vecchioli. "Mobilization of Iron in Water in the Magothy Aquifer During Long-Term Recharge with Tertiary-Treated Sewage, Bay Park, New York." *Journal Research U.S. Geological Survey.* Vol. 3, No. 1. pp. 93–98. 1975.

Ragone, S.E. and J. Vecchioli. "Chemical Interaction during Deep Well Recharge, Bay Park, New York." *Ground Water.* Vol. 13, No. 1. pp. Reprint. 1975.

Ramsey, R.H. III, J. Borreli and C.B. Fedler. "The Lubbock, Texas, Land Treatment System." *Irrigation Systems for the 21st Century, Portland, Oregon, 1987.* (American Society of Civil Engineers, New York, New York, pp. 352–361). 1987.

Razack, M., C. Drogue and M. Baitelem. "Impact of an Urban Area on the Hydrochemistry of a Shallow Groundwater (Alluvial Reservoir) Town of Narbonne, France." *Hydrological Processes and Water Management in Urban Areas, Duisburg, Federal Republic of Germany, 1988.* (International Hydrological Programme, UNESCO). 1988.

Rea, A.H. and J.D. Istok. "Groundwater Vulnerability to Contamination: A Literature Review." *Irrigation Systems for the 21st Century, Portland, Oregon, 1987.* (American Society of Civil Engineers, New York, New York, pp. 362–367). 1987.

Reichenbaugh, R.C. *Effects on Ground-Water Quality from Irrigating Pasture with Sewage Effluent Near Lakeland, Florida.* Geological Survey Water-Resources Investigations 76-108. U.S. Geological Survey, Tallahassee, Florida. 1977.

Reichenbaugh, R.C., D.P. Brown and C.L. Goetz. *Results of Testing Landspreading of Treated Municipal Wastewater at St. Petersburg, Florida.* Geological Survey Water-Resources Investigations Report 78-110. U.S. Geological Survey, Tallahassee, Florida. 1979.

Rein, D.A., G.M. Jamesson and R.A. Monteith. "Toxicity Effects of Alternative Disinfection Processes." *Water Environment Federation 65th Annual Conference & Exposition, New Orleans, Louisiana, 1992.* (Water Environment Federation, Alexandria, Virginia, pp. 461–470). 1992.

Rice, R.C., D.B. Jaynes and R.S. Bowman. "Preferential Flow of Solutes and Herbicide under Irrigated Fields." *Transactions of the American Society of Agricultural Engineers.* Vol. 34, No. 2. pp. 914–918. 1991.

Ritter, W.F., F.J. Humenik and R.W. Skaggs. "Irrigated Agriculture and Water Quality in the East." *Journal of Irrigation and Drainage Engineering.* Vol. 115, No. 5. pp. 807–821. 1989.

Ritter, W.F., R.W. Scarborough and E.M. Chirnside. "Nitrate Leaching Under Irrigation on Coastal Plain Soil." *Journal of Irrigation and Drainage Engineering.* Vol. 117, No. 4. pp. 490–502. 1991.

Robinson, J.H. and H.S. Snyder. "Golf Course Development Concerns in Coastal Zone Management." *Coastal Zone '91: Proceedings of the Seventh Symposium on Coastal and Ocean Management, Long Beach, California, 1991.* (American Society of Civil Engineers, New York, New York, pp. 431–442). 1991.

Rosenshein, J.S. and J.J. Hickey. "Storage of Treated Sewage Effluent and Storm Water in a Saline Aquifer, Pinellas Peninsula, Florida." *Ground Water.* Vol. 15, No. 4. pp. 284–293. 1977.

Sabatini, D.A. and T.A. Austin. "Adsorption, Desorption and Transport of Pesticides in Groundwater: A Critical Review." *Planning Now for Irrigation and Drainage in the 21st Century, Lincoln, Nebraska, 1988.* (American Society of Civil Engineers, New York, New York, pp. 571–579). 1988.

Sabol, G.V., H. Bouwer and P.J. Wierenga. "Irrigation Effects in Arizona and New Mexico." *Journal of Irrigation and Drainage Engineering.* Vol. 113, No. 1. pp. 30–57. 1987.

Salo, J.E., D. Harrison and E.M. Archibald. "Removing Contaminants by Groundwater Recharge Basins." *Journal of the American Water Works Association.* Vol. 78, No. 9. pp. 76–81. 1986.

Schiffer, D.M. *Effects of Three Highway-Runoff Detention Methods on Water Quality of the Surficial Aquifer System in Central Florida.* Geological Survey Water-Resources Investigations Report 88-4170. U.S. Geological Survey, Tallahassee, Florida. 1989.

Schmidt, K.D. and I. Sherman. "Effect of Irrigation on Groundwater Quality in California." *Journal of Irrigation and Drainage Engineering.* Vol. 113, No. 1. pp. 16–29. 1987.

Schneider, B.J., H.F.H. Ku and E.T. Oaksford. *Hydrologic Effects of Artificial-Recharge Experiments with Reclaimed Water at East Meadow, Long Island, New York.* Geological Survey Water Resources Investigations Report 85-4323. U.S. Geological Survey, Denver, Colorado. 1987.

Seaburn, G.E. and D.A. Aronson. *Influence of Recharge Basins on the Hydrology of Nassau and Suffolk Counties, Long Island, New York.* Geological Survey Water Supply Paper 2031. U.S. Government Printing Office, Washington, D.C. 1974.

Shirmohammadi, A. and W.G. Knisel. "Irrigated Agriculture and Water Quality in the South." *Journal of Irrigation and Drainage Engineering.* Vol. 115, No. 5. pp. 791–806. 1989.

Smith, S.O. and D.H. Myott. "Effects of Cesspool Discharge on Ground-Water Quality on Long Island, N.Y." *Journal of the American Water Works Association.* Vol. 67, No. 8. pp. 456–458. 1975.

Spalding, R.F. and L.A. Kitchen. "Nitrate in the Intermediate Vadose Zone beneath Irrigated Cropland." *Ground Water Monitoring Review.* Vol. 8, No. 2. pp. 89–95. 1988.

Squires, R.C., G.R. Groves and W.R. Johnston. "Economics of Selenium Removal from Drainage Water." *Journal of Irrigation and Drainage Engineering.* Vol. 115, No. 1. pp. 48–57. 1989.

Squires, R.C. and R. Johnston. "Selenium Removal—Can We Afford It?" *Irrigation Systems for the 21st Century, Portland, Oregon, 1987.* (American Society of Civil Engineers, New York, New York, pp. 455–467). 1987.

Steenhuis, T., R. Paulsen, T. Richard, W. Staubitz, M. Andreini and J. Surface. "Pesticide and Nitrate Movement under Conservation and Conventional Tilled Plots." *Planning Now for Irrigation and Drainage in the 21st Century, Lincoln, Nebraska, 1988.* (American Society of Civil Engineers, New York, New York, pp. 587–595). 1988.

Strutynski, B., R.E. Finger, S. Le and M. Lundt. "Pilot Scale Testing of Alternative Technologies for Meeting Effluent Reuse Criteria." *Water Environment Federation 65th Annual Conference & Exposition, New Orleans, Louisiana, 1992.* (Water Environment Federation, Alexandria, Virginia, pp. 69–79). 1992.

Tim, U.S. and S. Mostaghimi. "Model for Predicting Virus Movement through Soils." *Ground Water*. Vol. 29, No. 2. pp. 251–259. 1991.

Townley, J.A., S. Swanback and D. Andres. *Recharging a Potable Water Supply Aquifer with Reclaimed Wastewater in Cambria, California*. John Carollo Engineers, Walnut Creek, CA. 1992.

Treweek, G.P. "Pretreatment Processes for Groundwater Recharge." *Artificial Recharge of Groundwater*. Butterworth Publishers, Boston. pp. 205–248. 1985.

Troutman, D.E., E.M. Godsy, D.F. Goerlitz and G.G. Ehrlich. *Phenolic Contamination in the Sand-and-Gravel Aquifer from a Surface Impoundment of Wood Treatment Wastewaters, Pensacola, Florida*. Geological Survey Water-Resources Investigations Report 84-4230. U.S. Geological Survey, Tallahassee, Florida. 1984.

U.S. Environmental Protection Agency Office of Water, O.O.W.E.A.C. *Manual: Guidelines for Water Reuse*. EPA/625/R-92/004. U.S. Environmental Protection Agency, Washington, D.C. 1992.

Varuntanya, C.P. and D.R. Shafer. "Techniques for Fluoride Removal in Industrial Waste-waters." *Water Environment Federation 65th Annual Conference & Exposition, New Orleans, Louisiana, 1992*. (Water Environment Federation, Alexandria, Virginia, pp. 159–170). 1992.

Vaughn, J.M., E.F. Laudry, L.J. Baranosky, C.A. Beckwith, M.C. Dahl and N.C. Delihas. "Survey of Human Virus Occurrence in Wastewater-Recharged Groundwater on Long Island." *Applied and Environmental Microbiology*. Vol. 36, No. 1. pp. 47–51. 1978.

Vecchioli, J., G.G. Ehrlich, E.M. Godsy and C.A. Pascale. "Alterations in the Chemistry of an Industrial Waste Liquid Injected into Limestone Near Pensacola, Florida," *Hydrogeology of Karstic Terrains: Case Histories*. International Association of Hydrogeologists, UNESCO, Vol. 1. pp. 217–220. 1984.

Verdin, J., G. Lyford and L. Sims. *Application of Satellite Remote Sensing for Identification of Irrigated Lands in the Newlands Project*. 1987.

Waller, B.G., B. Howie and C.R. Causaras. *Effluent Migration from Septic Tank Systems in Two Different Lithologies, Broward County, Florida*. Geological Survey Water-Resources Investigations Report 87-4075. U.S. Geological Survey, Tallahassee, Florida. 1987.

Wanielista, M., J. Charba, J. Dietz, R.S. Lott and B. Russell. *Evaluation of the Stormwater Treatment Facilities at the Lake Angel Detention Pond, Orange County, Florida*. Report No. FL-ER-49-91. Florida Department of Transportation Environmental Office, Tallahassee, Florida. 1991.

Wellings, F.M. "Perspective on Risk of Waterborne Enteric Virus Infections." *Chemical and Biological Characterization of Sludges, Sediments, Dredge Spoils, and Drilling Muds. ASTM STP 976*. American Society for Testing and Materials, Philadelphia, Pennsylvania. pp. 257–264. 1988.

White, E.M. and J.N. Dornbush. "Soil Changes Caused by Municipal Wastewater Applications in Eastern South Dakota." *Water Resources Bulletin*. Vol. 24, No. 2. pp. 269–273. 1988.

Wilde, F.D. *Geochemistry and Factors Affecting Ground-Water Quality at Three Storm-Water-Management Sites in Maryland: Report of Investigations No. 59*. Department of Natural Resources, Maryland Geological Survey, Baltimore, Maryland. (Prepared in cooperation with the U.S. Department of the Interior Geological Survey, The Maryland Department of the Environment, and The Governor's Commission on Chesapeake Bay Initiatives). 1994.

Wilson, L.G., Osborn M.D., K.L. Olson, S.M. Maida and L.T. Katz. "The Ground Water Recharge and Pollution Potential of Dry Wells in Pima County, Arizona." *Ground Water Monitoring Review*. Vol. 10. pp. 114–121. 1990.

Wolff, J., J. Ebeling, A. Muller and H. Wacker. "Waste Water Irrigation Suited to the Environment as Shown by the Example of the 'Abwasserverband Wolfsburg'." *Hydrological*

Process and Water Management in Urban Areas, Duisburg, Federal Republic of Germany, 1988. (International Hydrological Programme, UNESCO). 1988.

Yurewicz, M.C. and J.C. Rosenau. *Effects on Ground Water of Spray Irrigation Using Treated Municipal Sewage Southwest of Tallahassee, Florida.* Geological Survey Water-Resources Investigation Report 86-4109. U.S. Geological Survey, Tallahassee, Florida. 1986.

INDEX